FLYING KNOW-HOW

FLYING KNOW-HOW

by Robert N. Buck

An Eleanor Friede Book
MACMILLAN PUBLISHING COMPANY
New York
COLLIER MACMILLAN PUBLISHERS
London

Macmillan Publishing Company
866 Third Avenue, New York, N.Y. 10022
Collier Macmillan Canada, Inc.

FIRST MACMILLAN EDITION 1987

Library of Congress Cataloging-in-Publication Data
Buck, Robert N., 1914-
Flying know-how.
"An Eleanor Friede book."
Includes index.
1. Airplanes—Piloting. I. Title.
TL710.B76   1987   629.132'521   85–19853
ISBN 0-02-518260-9

Macmillan books are available at special discounts for bulk purchases
for sales promotions, premiums, fund-raising, or educational use.
For details, contact:
Special Sales Director
Macmillan Publishing Company
866 Third Avenue
New York, N.Y. 10022

10 9 8 7 6 5 4 3 2 1

PRINTED IN THE UNITED STATES OF AMERICA

*Dedicated to those pilots with whom,
for thirty-six years, I was privileged
to share a seniority list, the sky,
and a conglomeration of airplanes from
Douglas DC-2s to Boeing 747s.*

My sincere appreciation to these kind folks
who helped in various ways:

ROBERT O. BUCK, CAPTAIN J. E. BURNHAM,
LEIGHTON COLLINS, ANN LOWE, JOHN MACONE,
CAPTAIN JACK ROBERTSON, FERRIS URBANOWSKI.

Contents

1

Traits of a Pilot

I've seen men and women from fourteen to seventy-five learn to fly. Fat people, skinny people, tall ones, short ones—all kinds. There isn't any outward image that signifies a good pilot, and no truth to the old picture of a tall, handsome, blue-eyed Adonis being the best pilot. The color of eyes or skin doesn't mean a thing.

When I was a new copilot, on one of my first trips, it was rather disturbing to see a corpulent old guy, a little bleary-eyed, puffing and obviously out of shape, struggle into the seat of a DC-2, let out the safety belt to get it around his middle, shorten the rudder pedals to fit his stubby legs, and then fly the airplane with the beauty of a ballet dancer and the wisdom of Solomon.

Appearance doesn't make the pilot, it's what's in his head

that does: what he knows and how he uses it. So we're going to explore what's in that head, what should be in it, how he thinks and acts, and the many, many tricks of the trade he's picked up through the years. What flying character he has developed. And we will talk a great deal about pilot character; really, it's what this book is about, and everything we talk about should be part of a pilot's character if he's a good pilot.

Flying takes little physical effort, but it takes a lot of thinking, response, and conviction.

The list is long but interesting. It runs from enthusiasm to being scared—which is necessary to some degree. Good pilots have to know what being alone means, how to use patience, pride, and humility. They have to train themselves to think and act certain ways because, no matter how much stuff the Federal Aviation Regulations (FARs), ground school, or an instructor drill into us, the final grasping of how it's done is up to the pilot himself. Actually, we're all self-taught; a good instructor only makes it easier for us. We have to teach ourselves, because the learning process never, ever ends.

Habits

Habit is a part of a pilot—good habits and bad ones—some to develop and some to get rid of. A pilot I flew copilot for, way back when, always reached down and unlocked the tail wheel of the DC-3 immediately after he touched on landing. It was almost a reflex action, it had become so much of a habit. It wasn't smart, because at high speed, before you could taxi gently with the brakes and throttles alone, the DC-3 might go catywompus in a vicious ground loop. Well, of course, he finally did. Luckily he was on a wide airport, and sailed around in a tight turn without hurting much except his dignity and the outside of a wheel and tire. But the next time I flew with him, I noticed he'd gotten over that habit of unlocking the tail wheel as soon as he touched down.

Look Me Over

Which shows the value of having someone look at your flying now and then to see if you've developed any bad habits you haven't noticed. Ideally we ought to look at ourselves objectively and try to see our own bad habits. It's another good pilot trait we'll talk over: honest self-inspection and appraisal.

Pilots have to learn about complacency, and the fact that it may be our worst enemy. And about sensory illusions, when you can't believe your eyes.

We have to understand emergencies: what they are, how to handle them, and the magic one-two-three moves to make automatically in the first split second when something goes wrong and we're apt to just sit there and do nothing—or move too fast.

We have to learn about thinking ahead, way ahead, because all flying is based on what's going to happen or be required next.

There are many more things that go into our make-up and character as pilots. Most of them are gotten through the thing called "experience." Now and then an experience is too much, and instead of being an incident to learn from, it does us in. With thinking we can avoid these catastrophes.

So what this book is trying to do is help us collect experience in advance, sitting in a comfortable chair at home rather than upside down at 10,000 feet. Of course it can only help, and if it helps even a little bit, it's worth it.

So let's dig in and see what makes us how and what we are . . . and ought to be.

ENTHUSIASM

Of the many traits that make up good pilot character, enthusiasm is high on the list. If a pilot isn't enthusiastic about flying, he doesn't belong in the air.

There is so much to learn and re-learn; there are new things constantly arriving on the scene that demand attention. If all this has to be done without enthusiasm, it is drudgery, and when something is drudgery, it isn't done well or completely.

I've talked with a number of TWA's check pilots, and I recall my own experiences of giving checks and working with new pilots as they upgraded to captain. We all agree that the pilots who fail almost all lack enthusiasm. To some pilots it's just a job, and it's these who have the highest failure rate.

Many years ago I became friendly with one of our captains—a good pilot who'd flown fighters in the Air Corps (it was called that when he graduated)—who had gotten a job with the airline when his Air Corps days were over. He was enthusiastic about flying fighters, because it was fun. He was enthusiastic about the airline at first, but he really didn't give a hoot about aviation or airplanes.

One Sunday I inveigled him into visiting our local airport with me. It didn't impress him a bit. I suggested he go around the patch with me in a Cub. He stood looking at me aghast. "Are you crazy? Me go up in that thing?" And shortly he said he was bored and let's get on home.

Well, as the years passed his flying got sloppier and sloppier. He was rough with the airplane. He checked out in 707s, but one day, after having flown them for a while, he failed an instrument check. He went back to copilot for a stretch, but, prodded by pride and economic loss, cranked himself up and got back to captain, flew a few more years, and retired.

There wasn't a thing wrong with his ability. He just didn't have enough enthusiasm to keep up and keep ahead of the business. He was happy to retire.

I flew a lot with a certain copilot—a nice guy who could fly well, but was interested in boats, not airplanes. He carried yachting magazines and boat information on trips, and was anxious to show you pictures and specifications. But he should have been carrying airplane and procedure manuals instead. He had a rough time trying to check out as captain, and didn't make it. It was exasperating to me to see him miss when I knew he could do the job if he really worked at it. But he started working too late, and without the enthusiasm he needed. Information wasn't at his fingertips; he wasn't sharp on the airplane and its systems. He couldn't learn in a few months the stuff he should have been getting over a few years. He was a good example of the fact that just knowing how to fly well doesn't make a pilot; there's much more to it than stick and rudder.

It's really more than enthusiasm, it's love of flying, the entire ball that makes up aviation. If you don't love it enough to be enthusiastic about it, don't fly—go do something else.

LAZINESS

The opposite of enthusiasm is laziness. Oh, I guess you can be enthusiastic about something and still be lazy, but if you're so lazy as to not get the job done, then you're not very enthusiastic. But laziness in flying isn't good. Not just the matter of being too lazy to get out the books, or poke around trying to learn more, but laziness in the air brings trouble.

Being too lazy to identify an omni each time one is tuned is dangerous, as is not draining sumps, bothering to chock the wheels, drawing a course line on a map, listening for weather.

Flying VFR (Visual Flight Rules) near a big airport, being lazy is not looking up approach- and departure-control frequencies to see how traffic's moving and maybe catch a jet taking off and being vectored your way. Laziness is not listening to the Automatic Terminal Information Service (ATIS) as you go by so, if you want to land suddenly, you know which way it's being done down there.

You can make your own list, but not being lazy, and doing extra things in flight and on the ground, make for more secure flying.

CURIOSITY

Wrapped up in this package of enthusiasm and laziness is curiosity.

You own an airplane, or rent one. Have you been curious enough to read the airplane manual? Curious enough to read it carefully and study the takeoff section and really learn it in relation to temperatures, altitudes, and the airports you use? Are you curious about the cruise power settings vs. range and altitude? Have you been curious enough to know exactly what the most efficient altitude is for the airplane? How long it takes to climb to various altitudes under different temperatures?

Are you curious enough to visit the nearest Air Traffic Control (ATC) center and learn what goes on (and ditto for an IFR [Instrument Flight Rule] room at a major terminal)?

There's a pretty long, never-ending list. We can see things to be curious about every day, enough to make a list that's impossible to finish checking off. But new things present themselves, and we have to be curious about them. There's always something we hear of and wish we knew. Curiosity adds interest to our flying and helps create enthusiasm, and that combination tends to kill laziness.

PRIDE

Pride goes into many decisions, and nothing can break your neck quicker than unswallowed pride. If you're to live and fly, you don't allow pride to enter a decision. You must develop a philosophy of not giving a damn about what people might think. And in matters of pride one generally finds that what he thought "people" might think wasn't important at all; "people" really didn't care enough about him to matter anyway!

Pride is deep in flying. Sometimes it should be called "showing off": You're too high and fast for a landing, and bust it up trying to get in rather than pour on the coal and go around. Or you've just broken out on an instrument approach to a slick runway. You're a little fast, but since you want to make a smooth landing and show off to the passengers, you hold it off and try to ace it on. The excess speed uses up a good part of the runway: When you do touch, there isn't enough braking to keep from sliding off the end. It would be better to bang it on a little hard, early on the runway, and not give a hoot if the landing is rough, as long as you have all the room possible to get stopped.

The funny part of pride, too, is that generally the showing off that goes with it is showing off to bystanders and people who don't know what it's all about anyway. Doing it the right way—going around again on an overshoot, staying within the airport once down, not running out of fuel—impresses the people in the business, your peers—and they, after all, are the ones you really want to impress!

Pride in one's work can be a big advantage, but pride mixed with ego is dangerous.

HUMILITY

The humility of knowing that the sky and its forces are bigger than any man, and that it's necessary to be humble enough to know we cannot always win. Humility before a big thunderstorm makes us say, "Okay, you win. I'll turn around, land, and be humble while you crash, blow, and send down deluges of water."

It's knowing that there is a point at which visibility is low enough so that we say, "I can't see enough to land, I'll pull up and go elsewhere."

It's the ability to say "I don't know." Somehow it often seems like a sin in aviation to admit you don't know something. There's a great challenge of one-upmanship. One-upmanship to others is one thing, but to one-up ourselves is disaster. Saying, "Oh sure, I know the fuel system on that 421," when really you don't—not in the intimate detail you should—is not being honest with yourself, or with the 421's fuel system. Humility brings the opportunity to use that wonderful, honest, and clean expression: "I don't know."

And humility means the wisdom to listen. There isn't any better training ground for information than bull sessions around airports. The only guy who doesn't gain from them is the one who isn't humble enough to listen, but has to do all the talking.

A great source of pleasure for me is sitting in an airport pilots' lounge, back in a corner, not part of things. I like to fake snoozing or reading a magazine while actually I listen, only listen. It brings many rewards. I never leave a session like that without a list of things to check on that the conversation suggested: perhaps something about FARs I didn't know, an airway procedure, a point in an airplane's opera-. tion—a lot of things. And all gleaned by listening, humbly.

RESPONSIBILITY

An obviously important part of a pilot's character, but what is the responsibility? How broad an area does it cover?

Naturally, we're responsible to our passengers, whether friends going along for the ride or paying passengers. Both are the pilot's responsibility. He's responsible for their safety, comfort, and accomplishment of what they want—a scheduled arrival, for example, or simply being delivered in good condition to where they want to go.

There are times when we stick our necks out, in weather or going across a bad piece of country on one engine. But if we're responsible, we don't do that with others. Taking risks is a solo operation, when there isn't anyone to hurt except ourselves.

But there's responsibility in that too, because of the effect our accident might have on the people around us, and on aviation.

Suppose someone decides, since he's alone and feeling frisky, to barge across some foul terrain. Then bad luck catches up, and the engine quits. He settles into a mess of trees way out in no man's land. It's his own fault, and tough. But now people begin looking for him. They stick their necks out, which people tend to do on rescue missions. Now his risky act hasn't affected just him; he's made a lot of other people risk their necks too.

Irresponsible acts like showing off around airports, low circling of houses, and all the rest can cause injury or property damage to people below. But it does other things. A new pilot looks up to a more experienced one. The experienced one does some silly showoff maneuvers, successfully. The new pilot says, "Man, that was pretty, I can't wait to try it!" and breaks his neck.

We never know how far and to what far corners acts of responsibility go. The world is such that there is almost no act possible that in some way doesn't affect another person. Simply stated, when we decide to fly we assume responsibility. It never leaves, and we must realize that the joys of flight will always be disciplined by responsibility: the responsibility of doing the best job possible.

SELF-CRITICISM

There's nothing new about this. The oracle of Apollo at Delphi said it: "Know thyself."

Perhaps self-criticism is one of the best ways to improve one's knowledge and ability in flying. There's ample opportunity.

I used to have a two-hour drive between airport and home when flying the Atlantic from Kennedy. A lot of that time on the way home after a trip was spent in going over the flight to recall how I could have done it better, where mistakes had been made.

I can honestly say I have never flown a flight in which there wasn't some point I thought I could have done better or safer. Sometimes the point might be small, but there always is one. If I said to myself, "There's nothing to criticize, it was a perfect flight," I'd wipe the thought from my mind and look deeper for errors, because, although I'm not a superstitious person, it seems that every time I begin to think I'm doing a great job, something goes wrong.

The fact is that flying has so many facets and is so complex, there isn't any way to get it all done with perfection. The effort is to keep the errors and omissions small and not catastrophic. That isn't too difficult with reasonable care. By making the small details as perfect as possible, we prevent big errors from occurring.

What sort of things would bother me? Well, why did I have five knots extra speed landing at Rome? I didn't need it.

Was 31,000 feet the right altitude out of Paris? For the stage length to Rome, 35,000 would have been more economical. The winds looked better at 31,000, but I really should have gotten the book out and checked it thoroughly. Lazy. Maybe I was tired because I'd just flown from New York. No, that's a weak alibi. It was just laziness.

I put the autopilot on at 10,000 feet out of New York. To keep in practice, I usually fly by hand until cruising altitude. I should have done it then.

The descent into New York: The ATC wanted me 17,000 feet 30 miles east of Hampton, which is too far from New York to be that low; it's a big waste of time and fuel. I shouldn't have argued so much, and maybe so nastily, with the controller; it took up air time and was undignified. I

call the airport and say he was returning. Concerned, I called and asked if he had any problem.

"Yes, damn it, I forgot to turn in the rental car!" Not really important, but in his mood he could have forgotten something that was.

Too Slow?

But we'd like not to confuse setting a proper pace with being too slow. Plodding, fussing, and fidgeting just use up time, and in flying we don't want to use time unnecessarily and excessively. If we're a lot slower than the next guy, it's time for self-appraisal and learning to speed up. There are too many times when getting the job done quickly is important.

You want to take off on a cross-country flight. A big thunderstorm/cold front is approaching; you're going the other way, and you've got to get off before it arrives. This is the time to get things done quickly. It's a great demonstration of patience and cool to get it done properly and still get off before the front. If we haven't trained ourselves to move quickly and efficiently, we may take too long and never get out, or, worse, take off when we shouldn't, with the front too close and ourselves in a stewing sweat and a state of some confusion.

You Can Go Faster

Training to speed up is a matter of method—that is, developing a routine plus clearing our mind of superfluous items. Being slow is often a lack of confidence. We go back and check and recheck because we're not sure we've already checked. This can go on to a painful degree. But if we develop a plan that's logical in the order of checking, then we check each item and, being sure it's checked, don't go back and do it again.

While we want to learn to do things without wasting time, we also need to realize when we cannot beat that cold front and it's time to put the airplane back in the barn and patiently wait for better weather.

It Can Do You In

Impatience has probably killed more people in weather than any other one thing. An airport is below limits but, impatient to get in, the pilot tries an approach—bingo! A pilot is on low approach and, at 400 feet, gets glimpses of the ground. Despite a 200-foot minimum and glide slope that will take him to touchdown, he's impatient to see the ground, so he shoves it down, below the glide slope. There's scud ahead; he flies into it and hits an obstruction. Plowing into a mountain in marginal VFR conditions because a pilot was impatient to get somewhere is classic. Make your own list, but any list will show that impatience and weather are a dynamite combination.

Patience makes engines run better; impatience can make them run worse. Taking off with the oil not quite warm, or the engine still too cold to take full power properly, is at best hard on the engine and at worst can mean a failed engine in a critical part of the flight.

Don't Con Yourself

Impatience in these areas amounts to a con job on ourselves. Say it's a cold winter day; the airplane has been sitting out all night and has a coating of frost. An impatient guy says, "It'll dissipate." Or "It won't bother the way she flies."

Well, often it will, and more than one airplane has finished as a pile of junk off the end of the runway because of frost on the wings and a pilot too eager to get going. Maybe a little lazy, too—too lazy to bother getting the frost off the wings. But impatience probably made him start that takeoff when he shouldn't have.

Aside from the actual danger using patience may avoid, it also makes for a more pleasant life in the sky. The air traffic system is where we all need Job's patience. Being held, awaiting clearance, sitting in a holding pattern, flying an altitude we don't like, or going a long route that's out of the way are all part of ATC. Letting it get to you will cause ulcers and difficult days. This doesn't mean we shouldn't try to change the system and make it better, but that's an "on the ground, see your Congressman" job. While we're in the air flying ATC, we have to be patient, and long-suffering.

Impatience often brings sharp words and argument on the air that use up time and get controllers uptight. (Often they could use more patience, too.) The result is a shambles. But a patient guy who keeps cool and calmly talks to the controller often gets a lot more done toward expediting his flight than the screamer does.

It Pays Off

I was sitting in Paris, ready to take off for New York. I called for clearance to start engines, which you must do at Paris and most overseas airports. The controller, a pleasant-sounding French gal, said there would be a delay of two hours because of a slowdown over a new work contract. I wasn't very happy at the thought of their making over 300 people sit for two hours, but there wasn't any point in getting nasty and uptight. So I responded as sweetly as possible, and tried to sound as though the slowdown was a great idea and all right by me. But I did add, at the end of the transmission, "If there's any way I can get an earlier departure, I'd appreciate it."

In about ten minutes she called and said, "There are no delays via Abbeville."

It wasn't the way I was flight-planned or wanted to go, but it would get me in the air and somewhere over England; then I could sort it out with the British. So I said I'd go that way, and did, saving a lot of time. Patience and sweet-talkin' a nice-sounding girl had won a good point. I'm sure that if I'd been impatient and short she could have locked me in for the full two hours.

Patient Pros

All the good pilots I know are patient people—I mean the really good ones, like Jim Gannett of Boeing, Dwane Wallace of Cessna, Al White, who tested the B-70 supersonic bomber, and many airline and general-aviation pilots. They don't scream, shout, and run about when things are difficult. They have that quiet, confidence-exuding quality of patience and self-control when everything around them is coming apart.

RESOURCEFULNESS

When things don't go perfectly, or something breaks down and you're suddenly without it, it's time to be resourceful and make something else do.

A good example is the fellow on instruments who loses his altimeter. What does he do? Well, if he can get at it, he'll disconnect the manifold pressure gauge and use that, because it, like an altimeter, reads in inches of mercury. An inch is about 950 feet—not exactly, but the lower you are, the more accurate an inch per 950 feet becomes. Of course, there isn't any altimeter setting; the gauge will be about standard setting. But if he remembers what the last altimeter setting was, he can add or subtract that amount to get more accuracy. If it was 30.00 inches even, that's 73 feet higher than 29.92—not much to worry about. Oh, he wouldn't want to shoot a 200-foot approach, but he could use it for terrain clearance and a letdown to, say, 1000 feet. Since .10 of an inch is about 95 feet altitude, he'd want to level off about 200 feet above the target altitude, with an altimeter setting of 29.70. Of course, you can't read a manifold pressure gauge that closely, but it does give a rough idea of what you can do. If a pilot's curiosity is high, he'll think about this some free day and look up the altimeter/ altitude relationships. It's kind of fun, and might be useful someday.

Of course, a better instance of resourcefulness in this case is the pilot who, doing a lot of instrument flying, decides it would be smart to have two altimeters anyway. An excellent idea!

Stars

I had curiosity and resourcefulness work for me when I was a youth of twenty-two trying to make a world's distance record for light planes. With my cousin, Lee Bellingrath, as copilot, we took off in a Lambert 90-horsepower Monocoupe from Burbank, California, dropped the landing gear, and headed east, hoping to break the world's distance record, which was held by two Frenchmen on a flight from Paris to North Africa.

In the middle of the night our compass died. Due to vibration in the airplane, caused by a cracked motor mount, the compass card jumped off the jewel and slopped over on its side, dead. It was a clear night with the moon coming up, although the ground wasn't discernible. I knew the moon came up in the east and that if I headed for it we'd be going east. Also, I'd done enough star study to know where the North Star is—right there where the Big Dipper pointers say. If you keep the North Star off your left shoulder, you've got to be going east. At dawn we were over Wichita, which is about only 50 miles off the great-circle course. From then on it was map reading in daylight, with the earth visible and understandable. We landed that afternoon at Columbus, Ohio, with a new world record.

Knowledge of the stars is handy, and interesting, for a pilot, even if he never plans to use celestial navigation.

Resourcefulness is part of judgment. Suppose we're on an overwater flight with sketchy navigation aids, the main one being an Automatic Direction Finder (ADF). We need bearings to plot a position fix. Tuning a station far off brings an ADF needle that swings through 30° or so. That isn't very accurate. But a resourceful pilot can make use of the information. He studies the swinging carefully and tries to get an average, then tunes another station, maybe not much better than the first one. But he averages those swings as best as possible and plots them. He has a fix. (He should apply all the corrections possible, such as convergence factor of the chart, variation, deviation on the compass, and ADF, if any.)

Averages Help

A resourceful pilot doesn't use that fix except as reference. He takes another fix and plots that, and another, so that after a half-hour period he has a string of fixes. They should show some consistency. Using the distance from first to last fix, he plots his flight-plan groundspeed and gets a rough groundspeed check. It may show slower or faster, but it gives a ballpark idea of how the fixes agree with groundspeed. He averages all this out and, using resourceful judgment, says, "I'm about here."

It's often enough to keep near track and get a usable estimate, tempered by the realization that it isn't a perfect position and therefore shouldn't be used to the precise limits of fuel range, terrain clearance for a coming coastline, etc., that a precise fix would. Resourcefully, and with judgment, these factors are used within their limits. But our pilot just didn't give up when he first tuned the ADF and saw its erratic wanderings.

RESPECT

Not only for one's elders—here we mean equipment. If we don't respect our equipment, then we don't take care of it, and if we don't take care of it, failures occur.

Respect, too, for limitations, both in numbers in a manual we're not supposed to exceed, and those put on a pilot by himself or authorities.

Respect for a number is the lowest minimum an approach plate says you can descend to. Famous last words include "I know this approach, you can shove her down another hundred!" People say things like that; I've heard 'em. But if we say that, we really don't know the approach, because when that minimum number was put in the book it was carefully worked out as the lowest safe minimum, and it takes into consideration all obstructions, airport, lights, and so on. Nearby is Knapp State Airport, Montpelier, Vermont. It's got an NDB (Non Directional Beacon) approach from the south, minimum 863 feet above the field. I don't cheat an inch on it, because a little to the left, on final, is a big hill. Looking at it VFR, it's startlingly close. You'd have to be stupid to cheat there.

Respect for the airplane numbers such as max speeds,

gross weights, and all the rest is important because the numbers mean something. The numbers that go into a manual are worked out by the factory making the airplane and the FAA. The numbers aren't capricious, they are carefully considered.

The engineering department of a factory goes over its data; then test pilots, who can fly that airplane better than we ever will be able to, and know more about it, check out much of the engineering information and introduce new information gained from their testing of the airplane, including taking it over and under its design numbers.

Then the test pilots, both factory and FAA, plus the engineering data men, both factory and FAA, work to come up with the airplane's final numbers, which are put in the manual for us to respect.

Naturally, the factory tries to get the most performance possible; it sells the product. But in working with factories, and flying with their test pilots on test work, I've gained great respect for the test pilot's interest in the man who will eventually fly the airplane in service. The test pilot tries to present him with the most useful and safest numbers he thinks the airplane can consistently produce. But we all should remember that the test pilot flies better than we do, and the airplane gets older, so it behooves us to use the numbers conservatively.

After years of association with the system, I firmly believe that the best numbers are the ones in the manual, and to think that I can make them better, exceed them, or outdo them is ridiculous. Challenging the numbers takes extensive equipment and careful test procedures that the everyday working pilot just doesn't have at his command.

Spit and Polish

Respect for an airplane means keeping it in clean, ship-shape condition. An engine that's dirty with oil is a fire hazard. Dirty windshields reduce visibility, and a dirty windshield when flying into an afternoon sun is almost a zero-visibility condition. Dirty, dented airplanes fly slower.

Respect is reflected in care during operation. Going easy on brakes, not slamming the airplane around. Not jamming the throttle open, but easing it in. There are lots of turbo supercharged engines now, and a sure way to damage one of them is by slamming in the throttle on a cold day. It's easy to overboost, as thick oil and cold moving parts don't respond quickly.

We're supposed to love our airplanes. Thousands of words have been written on the euphoria that exists between airplane and pilot. Well, the best way to show it is by respect . . . and maybe a gentle kiss on the spinner now and then.

PRUDENCE

It means, they say, careful management, economy, caution, and wisdom in the conduct of affairs. Which is a pretty good description of a proper pilot.

Careful management is preparation of our flight: weather, airplane, flight plan, all of it.

Economy is part of our frugal use of fuel and the airplane —not to save energy and money necessarily, but economy of fuel so we'll have enough to get there, and frugality with the airplane so it is in proper condition and ready to answer our demands.

Careful

Caution is suppression of the foolish. Proper flying brings its caution automatically. Being uncautious is the thing to be wary of.

Wisdom—ah, a big word. But isn't it the knowledge of everything there is to know about flying, and the use of common sense? Thoreau said it: "It's a characteristic of wisdom not to do desperate things." Could anything apply better to flying?

Complacency

Complacency is something I talk about all through this book. Without a doubt, it's one of the most serious problems in flying, and it becomes worse the more sophisticated an airplane becomes. If we're flying a difficult, unstable airplane, with an engine prone to failure and the weather bad, we aren't apt to be complacent. But fly a Boeing 747, for example—which is easy to fly, has a flawless navigation system and lots of range, plus the ability to land with 1200 feet visibility—and you've got something it's easy to become complacent with.

Modern general-aviation airplanes invite complacency. The engines rarely quit, and when they do an investigation generally finds that the pilot was too complacent with his fuel system in some way. These small airplanes have wide, tricycle landing gears that reduce the chances of ground loops, and can be driven onto the ground, in a thing called a landing, with little skill. The radios lead one to where he wants to go, even if he doesn't know where he is in between. This slickness and reliability invite complacency; they lull the lazy into thinking everything is always going to be fine.

But it isn't. And letting all these goodies take the place of thought and skill means, finally, that the airplane is flying the pilot rather than the other way around. This should never be, because someday the airplane will run out of brains—it really doesn't have many—and the pilot is going to have to use his own. If he isn't knowledgeable and up to speed in practiced skills, then there will be two dummies in the sky: the airplane and the pilot. Disaster surely will follow.

No matter how modern, sophisticated, and reliable the airplane is, the law of gravity hasn't been repealed, and each flight must be flown seriously, with all skills fine-tuned and practiced to their sharpest.

When a pilot says, "It's okay, I checked it yesterday, let's go," he has allowed a complacent attitude to enter his flying, and someday it will do him in.

Flying is safe, easy, and happy so long as we work hard at it. But if we are complacent and lazy, it will suddenly slap us down one day, and not be pleasant about doing it, either.

Part of a pilot's check on himself is to say, "Am I getting complacent?" If he asks the question, he probably isn't, which is a very good thing.

BOLDNESS

But despite all the cautions we've talked about, a good pilot must be bold. He must be bold to take off into a near-zero condition with thunderstorms en route. He must be bold to make that first solo cross-country flight. He must be bold to leave a friendly shore and point his airplane toward an island 2400 miles away with nothing but water in between.

Boldness is one of the fascinations of flying. The need for boldness excites; it gives a feeling of identity, of doing something different from man's daily humdrum.

Yes, we need boldness, but with it the other traits that make good pilot character. In that way, boldness doesn't become recklessness, something it's often confused with.

Perhaps the most perfect example of boldness was Lindbergh's flight to Paris in 1927. He certainly had to be bold, but for his day and time Lindbergh was also the pilot best prepared to make the attempt. His background as a mail pilot certainly gave him more weather experience than any others who tried. He used the best engine of the day, he had an airplane designed and built to the most up-to-date standards, and then he put everything in it the state of the art had that was useful for his flight. He used an Earth Inductor Compass, which was the flux-gate-type compass of that day. He tested the airplane at the factory and on a one-stop transcontinental flight. He studied the weather forecasts carefully and plotted his navigation precisely. Then, equipped as completely as technology would allow in both airplane and man, he boldly took off for Paris. Yes, he took risks we don't take in normal flying, but he had the character and class to be the best, and he was.

There *are* old bold pilots, but they have the rest of it, too.

2

Safety

"Safety" is a word we live by in aviation, but I sometimes wonder if it wouldn't be better if we never used it.

After all, everything we do in flying is directed at safety.
But the moment someone says something is a matter of
safety, it takes on a special aura, and that really shouldn't
be. If something is said to be safe or unsafe, special attention
is paid to that item. Well, what about all the rest? Isn't
everything we do a matter of safety?

You call for weather before a flight—that's to make it safe;
you carry the right maps and charts—that's to make it safe;
you take sandwiches and coffee along to keep you sustained
and awake—that's for safety; you preflight it—and that's for
safety. It goes on and on. So why pick out one item? It's all

for safety. I'd rather just call it "good operating practice" and forget this word "safety."

I have a feeling that when a particular safety item is over-emphasized, people get a sort of spooked feeling, and perhaps pay too much attention to that at the price of overlooking something else.

Another thing about doing away with the words "safe" and "safety" is that we might get a lot more done toward safety, because, believe it or not, safety is a difficult item to sell.

I've spent a lot of time in the world of safety, worked for it with the Air Line Pilots Association (ALPA), various manufacturers, and the FAA, and appeared before Congressional committees. One thing I've learned is that people give safety a lot of lip service, but when it comes to spending dollars for it, they easily rationalize why it isn't needed.

For many years the Air Line Pilots Association has fought for safety items. Sadly, many of the items they've requested the FAA to create and enforce, in the form of regulations or equipment and changes in airplanes, have come only after a serious accident demonstrated their need.

Long Time Coming

Safety takes a long time. In a report I wrote for the Administrator of the FAA in 1962, I suggested that Visual Approach Slope Indicators (VASI) be installed on instrument runways. The FAA's policy then was that if an ILS, with its glide slope, was on an instrument runway, the visual aid wasn't needed. It was short thinking indeed, because while the glide slope is okay when on instruments, you cease seeing it the minute you look up and out, trying visually to find the runway pavement for a graceful landing. And that isn't easy on a dark and stormy night. Well, after numerous accidents the National Transportation Safety Board, in 1974, recommended that VASI be on all instrument runways. It took only twelve years and a few accidents to get the point across!

One Way

Years ago we got a new type of Constellation that needed an alternate air source to the engine because we found in service that the engines iced under certain conditions. Without it we had real problems in wet snow, with erratic engine surging and even, on occasion, quitting. It took a lot of fancy flight-engineer work, plus altitude changes and avoiding weather, to keep things going. But we had a tough time convincing the company they should make the necessary modification to get the alternate air installed. It was expensive, and the aircraft would be out of service.

Well, one night I had a certain company executive on board, and I was back in the cabin talking to him as we ran through wet snow off the Nova Scotia coast. The engines began running irregularly, with fits of surging.

"What's that?" he asked.

"Oh, we're getting some engine ice."

"Why don't they clear it up?"

"Well, we really can't. You know, without that alternate air there isn't much to do."

"Don't you think you ought to be up front?"

I did, and I was anxious to get up there, but I kept my anxiety inside and said, "Well, there isn't anything I can do. That's a good flight engineer, and he's doin' all he can, but he can't cure it without that alternate air."

"There must be something you can do. I think you ought to be up there."

"Well, there's nothin' I can do."

"Bob," he said, "I order you to go up front."

"Okay," I said as I nonchalantly sauntered back to the cockpit, but I was very glad to get there! Shortly we flew out of the condition, but I was sure to let them run rough long enough to impress our executive. I must have, because it wasn't long after that that the order went out to complete the alternate air mod to the fleet.

Whose Fault?

The fact that safety is difficult to sell when it's expensive is one of the reasons accidents are generally called "pilot error." It's a lot easier to blame a pilot than it is to say that a major change should have been made on equipment.

There's that classic case where a certain wartime airplane was plagued with the gear being raised after landing. Pilot after pilot was blamed for being stupid and raising the gear instead of the flaps. Finally the idea dawned on someone to separate the gear and flap levers so it wouldn't be easy to make the mistake. *Voilà!* Problem cured. Of course it was a design error, but it also was an expensive modification to correct it. The original solution, which hadn't worked, was to drill the pilots not to make the mistake, when actually the airplane was at fault.

If we forget safety as a separate item and just admit that it's all simply good operational practice, we have a better chance of proving that changes we want are necessary to operation, and an economic need. With that approach, we might get a lot more done.

If one wants a change made, he has to justify it economically—prove that the expenditure will result in returned revenue. That goes for safety items too. So it just seems logical that if we tie everything to economics, we'll get the job done faster.

"Experts"

A nervous pilot seems, at times, to hide behind safety. He's afraid to do something or isn't properly equipped, so he says it isn't safe and doesn't fly. "You could get killed!" he says. It's an expression I often hear, and it riles me no end. Dammit! If a person feels that uptight about flying, he shouldn't fly! He doesn't understand what it's all about; he doesn't realize that it's a safe business if one learns the skills and techniques and absorbs the necessary information. With those he can set up limits of operation to fit him and the state of the art. Then he flies without worry.

Arm waving and saying, "They'll splatter an airplane all over the countryside and kill people!" indicates a pilot who uses emotion instead of intelligence to do the job. This guy doesn't belong in the sky. He's better off sitting home watching thrill shows on TV.

While I don't want to deprecate any of the honest people who work on safety and do an outstanding job—like Jerry Lederer of Flight Safety Foundation, Homer Mouden of Braniff and Eastern, Paul Soderlind of Northwest, Ted Linnert of ALPA, and a lot of other guys—I still can't help but mention the screwballs who get in the safety act.

There are two basic types. One is the nutty inventor who always has the panacea for everything, even though he doesn't understand what everything is, or what it's about. I've suffered these types calling in the middle of the night and circuitously talking for hours. You cannot explain that, while you don't want to discourage them, they ought to go out and learn the facts of life before they blast off.

The other type is the pilot who, afraid to fly and ignorant of what really goes on, becomes active in safety matters—it's easy to get on a committee. This guy is closely related to the screwball inventor. He gets wild ideas that aren't based on anything intelligent. By exposure, and an ego willing to expound to an ignorant press, he suddenly becomes an "expert." He then influences others who don't know he's an idiot. He's a menace to safety, actually, and, hopefully, finally destroys his case by arm waving and emotional action. Unfortunately, some of this type get loose and never are "de-experted."

As we said, these types are basically afraid of flying, and their fear generally is based on ignorance. Often the safety "expert" is a dangerous guy in the sky.

That's one reason that makes me want to forget the word "safety" and turn it all into "good operational practice."

How They Happen

Only 10% of all accidents are accidental. What are the other 90%? Acts of omission or commission by people involved. That's pretty important, and ought to sink in. Summed up: It isn't the airplanes, it's the people using them.

Human failure can manifest itself as inattention, distraction, haste, preoccupation, anxiety, anger, fear, hate, guilt, frustration, and probably more—which says that we shouldn't fly unless we're emotionally in a pretty good state. That precludes fights with wife or girl friend. The truth seems to be that we can fly without accident if we control and police ourselves.

The points that catch my eye are inattention, distraction, haste, fear, and preoccupation.

Some of these relate right to the job at hand. Flying an airplane is a multi-action job. A pilot has a lot of things to think about, and do, yet many of them aren't directly related to one another.

Can You Juggle?

For instance, we're on instruments, so a busy job of scanning is required, keeping all the needles where they should be. At the same time, we have to be alert to what ATC is saying, and we have to be conscious of the airplane's condition—meaning, how is the engine, is the power set properly, is there carburetor ice, and all that? How's the fuel for the flight, and what about the weather now and for our arrival . . . alternates? So it's quite a juggling act, with a lot of balls in the air. The prime thing I've always noticed about jugglers is that they concentrate on their job, they aren't inattentive, you cannot distract them, they aren't preoccupied, and certainly they cannot hurry things beyond the natural rhythm of the balls going up and down. They may be scared and afraid, but it's under control. And, as oversimple as it may sound, those are the qualities a good pilot needs. If he's got 'em working, he doesn't drop any balls, which is a way of saying he doesn't have an accident.

Sum It Up

What we're trying to say here is that safety shouldn't make unnecessary work in flying; that being overhipped on safety is a danger sign which says a pilot may be headed for trouble, and that generally accidents are caused, they don't happen.

Safety too often becomes confused with emotion, and flying emotionally isn't the way to do it.

The best pilot is the one who knows his craft thoroughly and, armed with this knowledge and skill, does his flying calmly, with enjoyment, and safely, from start-up to shutdown, without ever thinking, or saying, "safety."

3

Fear

Fear, as someone once said, kills the mind. And that's the key; we cannot allow it to kill the mind and stop our thinking.

We all experience fear at one time or another. To deny it is to kid ourselves.

Fear comes in different shapes and forms. There's fright: That's quick, like when we're flying peacefully and suddenly the engine misses in solid, unmistakable fashion. There's that quick jab of fear, and the adrenalin starts to move. Adrenalin helps us to move faster; it's a physical aid during fear to help us run away. This isn't especially good in flying, because, strapped to a seat, in the sky, there isn't any place to run! So we sit and take it. But, running or sitting, fright is a quick jab. Fright can turn into panic and terror, which are violent and paralyzing.

Dread is another kind of fear. It's long-term, and is fear of impending things. Taking off into icing conditions we're not certain about, or into a sky full of thunderstorms, can have a certain amount of dread attached to it.

A big portion of any fear is the unknown: We don't know what's going to happen. It's like the engine that misses and shakes—we know there's something wrong, but the thing that scares us is that we don't know what's going to happen next. Is it going to quit cold and force a power-off landing? If so, where? We look carefully at the terrain, which might not be so nice because we were sneaking across some unfriendly country to save a few minutes. But just because the engine is getting rough and there aren't any good fields below, we need not panic; it doesn't help.

If the Worst Happens

Now, suppose the engine does quit cold and there's nothing under but craggy mountains. Such a situation can lead to terror, which we sort of visualize as a state in which all thinking has stopped and we stare ahead, wild and wide-eyed, as they do in horror movies. Even in the most desperate country, we can lessen the damage if we crack it up under control while still thinking and flying the best we know how.

Well, all this sounds desperate and, of course, rarely ever happens. It doesn't ever have to, really, and these states are caused by our own minds, attitudes, knowledge, and preparation.

People Are Different

Some people scare easily and some don't. I can remember having a little trouble in a DC-4 when one crew member went bananas and the other remained cool.

We were climbing out of Milan, Italy, in rain, on instruments. The ceilings below were about 400 feet, there were big mountains to the north, and we were being careful of our navigation, as you always are under those conditions. It wasn't any special thing, and we were going to Rome, where it was much better. We weren't even heavy. But I suppose it was slightly creepy if you were a nervous person basically. You certainly can do a lot with your mind when it's raining, you're on solid instruments, and the Alps are not very far away.

My flight engineer was a nervous guy. To counter his flying apprehension, he'd become very serious about religion. Now, I don't knock religion, but I've noticed that every time a flight-crew member gets overly religious, he seems to have a problem with being scared in airplanes, and his behavior may become erratic.

Well, anyway, about 4000 feet in our climb, number three engine got rough and jumped around in a pretty wild fashion. My God-fearing engineer went slightly nuts. He was under control, but very vocal. "We've lost an engine, God, we've lost an engine!" And he started to grab at knobs and switches to try to correct the problem, in what was a too-fast, jumpy manner. He was in such an emotional state, I feared he'd grab the wrong thing and do more harm than good.

My copilot, Lew Cook, a very capable individual, didn't see what all the excitement was about. He realized that we could shut it down and then either get back to Milan, even with 400 feet and rain, or, being light, paddle on down to Rome.

The DC-4 was pretty easy to manage with minimum crew, so I said, loud and quick, "Lew, you keep Frank under control, and I'll take care of the airplane."

Lew didn't use force, but he persuaded the engineer to calm down and quit grabbing at things while I got the power reduced, which helped the problem. We went on to Rome.

What's the Difference?

Now, what makes one guy cool and the other not? Good question with no pat answers. But there are things to think about.

Before answering, we might look at opposites of fear. Two of them are confidence and assurance. In flying, what does that all boil down to? Simply, it is being prepared— prepared by knowing the subject and having skills developed and practiced to the point where one can handle situations with confidence and assurance.

It goes back to knowing the airplane inside and out, knowing the air traffic control system, and knowing weather.

Of these, weather is the only one that has vague areas. We

can learn the airplane well; that's a matter of study. The same goes for the air traffic system. But weather has its capricious times, and we cannot count on its doing what it's supposed to 100%. But should this put us in a position to become fearful about weather? No! Even the new pilot can handle weather in such a way that it shouldn't frighten him.

Let's go back to the experienced pilot. He knows how to fly instruments well, can shoot a Category II approach to 100 feet with no sweat. He's got radar and knows how to use it. The airplane is loaded with anti-ice equipment. So he doesn't have a problem: He can handle ice, miss the thunderstorms, and get in with a very low ceiling. He's been smart enough to carry sufficient fuel to go to an alternate in case he cannot get in even with a 100-foot ceiling.

So what's he got to be scared about? Not much. If, however, he's making an approach and an engine quits on his two-engined airplane, he's going to gulp and feel scared. But he also will know the procedures to fly it on one engine, and he'll do the best job possible.

Now to the other end of the scale: a new pilot flying a Cessna 172 with minimum equipment, and with no instrument rating. Should he be any more scared than the experienced pilot? Not really, although I'm not silly enough to say this pilot will not have some apprehension, if only because things are so new to him.

But he shouldn't have fear if he plans his flying the same way our experienced, well-equipped pilot does. First, to know his equipment, what its systems are, how fast it cruises, and, very important, its range. And know how to fly it well by lots of practice. Also to know, by study, the navigation system and how to use it.

Weather is his problem, but he can remain relaxed and enjoy flying by realizing that there are limitations. The first one is that he doesn't have an instrument rating and therefore must not allow himself to get into conditions that require instrument flight.

We cause fear by taking chances that create nervous apprehension or even scare the bejesus out of us. So if our low-experience-level pilot stays out of clouds and doesn't take chances with weather, he reduces the fear factor. He's got to stay away from reduced visibility, precipitation, and low ceilings. For openers, he should give himself ample ceiling and visibility to fly VFR without straining to see where he's going. And he should always fly toward good or improving weather—never, ever, into deteriorating weather, especially toward darkness. This means two things. One is an adequate weather briefing, with particular attention to forecasts for steady or improving weather along the way. The second is looking out the window and seeing if the actual weather is staying good enough. If it isn't, he lands or turns around—which demands that he keep finger on map and know where he is all the time, and where the nearest airport to run for is.

It's also wise to listen for weather reports. But for the VFR pilot, looking out the window is the thing. If it gets poor and he cannot see comfortably and fly high enough to be well above the terrain, then he turns and goes back to good weather, or lands. He doesn't believe the weather anymore. It is tremendously important to remember that what you see through the windshield is what the weather is, and that supersedes the forecast! The forecast may say it's going to be good, but if you're flying in lousy weather there isn't any point in thinking, "The forecast says it's going to be okay, so even though it isn't, I'll go along some more because it should improve." Oh, how wrong! The weather is what you see, and if you cannot handle it, then get out.

No Fear

If our low-time pilot can realistically look at the weather and fly it as he sees it, within the area that is comfortable, then he will not get in a position that's fearful.

That fear is ignorance applies especially to flying. It isn't only a matter of knowing equipment, techniques, and skill, but knowing one's limitations and flying within them.

It's difficult to say what our limitations are. I don't know mine in a decimal-point, spelled-out way. But if I look at a hot new supersonic fighter, I know I wouldn't climb in and fly it . . . as much as I'd love to . . . because it isn't within my limitations.

As scientific as flying is, it also has its intuition. You just damned well know by feel, hunch, or whatever that something you're considering doing (or not doing) isn't smart, and if you don't want to scare yourself, or worse, you back off and let it go by without attempting it.

I don't know how you gauge intuition, but I like to remember the great golf pro Lloyd Mangrum, who explained playing what he called "percentage golf." Simply, he looked at a difficult shot and then tried to figure what his chances would be of bringing it off. Could he do it 8 times out of 10? Deciding what his percentage chance of successfully making the shot was, and then how badly he needed it for score, he'd try it or, if the risk was too great, make a safer shot so he'd lose only one stroke, rather than the two or more he might if the chancy shot didn't work.

In flying we normally are looking for a 100% chance on anything we try. If a pilot looks at weather, or sneaking through a mountain pass, or doing a foolish aerobatic maneuver close to the ground, and it isn't a 100% performance without risk of failure, he's sticking his neck out in a position where it can be broken. Does a pilot really want that? Well, we can play percentage chance-taking—and if it isn't 100%, be prepared to get scared.

In our DC-4 situation, where I had one scared crew member and the other quite calm, what was the difference? They were both knowledgeable men in their trade. So why was the engineer scared? Well, he knew his subject, but he wasn't a pilot and didn't know, or really understand, how serious or not serious the engine failure was in relation to our total flight picture. So in a way his fright was lack of knowledge or understanding.

But there are cases where two people with the same knowledge and background have different fear levels. Why? It beats me, but man's emotions go back so far in time, are affected by such complexities as inheritance, environment, incidents, and other things, that the reason one man becomes more scared than another is confusing and difficult to figure out.

But no matter why, or what, it's certain that knowledge and the feeling that one's skills are developed to their utmost level will give one courage to get through a tough situation.

There is also the fact, and it is a fact, that, done within normal restraints and pilot ability, flying is very safe. I can look back over forty years and truthfully say that the serious problems I got into (thankfully, they were few) were mostly my own doing. Yes, I've had emergencies—engines quit, fires, hydraulic failures, electrical, and others that didn't have anything to do with how I'd flown the airplane—but being prepared and knowledgeable always made the situation controllable and the landing successful.

So what does that leave us? There is the rare catastrophic failure, such as a turbine wheel burst on a jet engine that may tear the structure apart, or possibly an event outside the pilot's control, such as an ATC error that caused a collision on instruments. But the record shows odds for such cases to be infinitesimal. Which says, again and again, that most accidents are caused, they don't happen.

Where does that leave us in relation to fear? It says we can and must subdue fear. We cannot shut off being scared, but if we have knowledge and skill we can keep thinking and working, using that knowledge and skill, even if our knees are shaking and palms sweaty, and even if an event is catastrophic—you never know what may show up at the last moment and save the day.

I admire people who can keep thinking, act cool, and show a calm appearance even though they are scared inside. That's real courage . . . my kind of folks. They don't allow fear to kill their minds.

4

Smooth Flight

A smooth pilot is a good pilot, but not just because it feels very comfortable to fly with a smooth pilot, or that he's much easier on the equipment. The important point is that in being smooth he does a better job of flying.

If a pilot is jerky, he tends to overcontrol, which leads to further jerks as he tries to recover from the overcontrol. Flying a sensitive airplane that may have dutch-roll characteristics will not be easy for the ham-handed pilot. A delicate instrument approach will be more difficult for the rough pilot.

Most flying is a matter of changing attitude, making small changes, and then stopping them exactly where one wants. It is easy to fall into the habit of watching, say, a change in attitude on the horizon, artificial or real, and then, when the desired attitude is reached, jamming the control the other way to stop the airplane from moving any farther. If we're climbing, such a fast stop will bring a brief feeling of weightlessness to passengers; a pullup will sit them down in the seat with some G's. It is difficult to hit the exact attitude wanted by a quick jab of the controls. Over- and undershoots occur, making for not only a rough feel, but a poor job of precision flying. During a tight ILS this can lead to localizer and glide slope wanderings, and sometimes a missed approach.

With smooth anticipation we approach the attitude desired and then begin to stop it smoothly with some lead. This can be done in little amounts and big amounts. It takes practice and good scanning.

Being too smooth—that is, moving too slowly to correct—can cause overshoots too, but this is prevented by good scanning so that an attitude never gets far out of line before it's stopped and corrected. If excursions of attitude are kept small, then corrections are small and the entire flight path is smooth and accurate.

Keep It Small

I was flying in the Concorde simulator with Gilbert Defer, one of the Concorde's test pilots. I wanted him to demonstrate an ILS using hand throttle to keep the proper airspeed. That's difficult in a delta-wing airplane. To ease pilot workload, most all the Concorde landings are made with automatic throttle. You simply dial the approach speed wanted—say, 160—and the computerized autothrottle gives you exactly that all the way down the slot. What I wanted to see, and later try out myself, was how difficult it would be if the autothrottle failed. So we were making ILSs with auto-throttle off.

I asked Defer to do one before I tried. The approach speed we used was 165 knots—and the airspeed never left 165! He flew it beautifully. But the interesting thing was that you hardly saw him move the controls; his motions were almost unnoticeable. He wasn't using shoves and pulls; he was simply applying pressure when he needed it. His scan was so good that he would catch an excursion as soon as it started; then he would apply just a little pressure in the direction needed to correct it. And mind you, he was doing the localizer and the rest of the ILS too. "It is a matter of small pressures and then releasing them before anything gets big." He meant before the airplane wanders off in big amounts. He was super smooth, as all good pilots are.

Stop It Before It Starts

A spiral dive is the classic way most airplanes are torn up with lost control on instruments by inexperienced pilots. Well, how does a spiral dive go? The wing drops, and the airplane begins to turn. Then the nose drops slightly, and the turning flight becomes descending. If the wing drop were stopped by small aileron pressure when it started, the spiral dive never would get going. But if the turn continues and the nose gets down, the speed builds, and a bigger correction is needed to stop the thing. Now, in the low-experience pilot's mind that increasing airspeed becomes the paramount thing; he may panic to stop that increasing airspeed at all costs. If he pulls back he's doomed: That tightens the spiral dive, as anyone knows. The experienced pilot gets the turn stopped and wings toward level as he stops the dive. Of course, he does it together, and almost in one action. If he's in a big hairy dive, with airspeed up to red line, he had better be smooth! If he isn't, he'll haul the wings off. It's better to be smooth and gradually pull out, even getting a little more speed in the process, than to jerk back and try to stop the dive. That jerk will create big G's, which we don't want and the wings cannot stand.

The real point is that, with a thorough scanning pilot who moves in smooth, small *pressures,* the wing will not get down very far at all, and the spiral dive never will start—which is one difference between a good instrument pilot and a poor one, or one who cannot fly instruments at all. But, either type, it pays to be smooth.

Smooth All Over

Smoothness begins with starting the engines. You see engines started with a great roar and then a quick jerk of the throttle to get them down to reasonable idling speed. That's hard on them. How much nicer to see a quick but smooth power reduction.

The use of brakes in taxiing is appalling to watch at times: jerky action, the nose bobbing up and down, and sudden stops until you wonder how the structure takes it.

Using brakes smoothly is a matter of anticipation—thinking ahead to determine the path of the airplane and when brakes will be needed—and then smooth application. Brakes, as it's been said by experts, are really an emergency device.

Another area of misconception is that brakes are part of the landing-distance requirement of the airplane. Certainly, the airplane is certified for field lengths using brakes, but the awful sight of someone touching down halfway up the field and then smoking the brakes to get her stopped is shocking indeed. At the average airport brakes need only the smallest, most tender application to get the airplane stopped and turned off. Our local airport at Warren-Sugarbush, Vermont, is 2600 feet, with a turnoff about halfway down. The average general-aviation single, flown by a good pilot, touches near the runway's beginning (there aren't any obstructions) and turns off in the middle without using much, and often no, brakes at all. I've even landed a Cessna Citation there, and turned off in the middle with only moderate braking. Yet by poor flying I've seen Cessna 172s pass that turnoff still in the air, finally land, and then pour on brake to get stopped within the field. Ye gads, what are people thinking! Someday those brakes, beat up from misuse, will fail and that 172 will go off the end.

Big or Small

Big airplanes aren't any different. I had an engine fail because of oil loss on a 747. I'd taken off from Los Angeles and was headed for London, so we were heavy. About Bryce Canyon the engine packed up, and we shut it down and headed back to LAX. At 630,000 pounds, we were well over the legal gross landing weight (which for that 747 was

564,000 pounds), but the charts all said that we could land safely within the airport. Well, we try to save fuel, and it was a nice day, and we were okay by the charts, so we decided to land as we were and not dump 11,600 gallons of kerosene in the Pacific Ocean.

So we landed, no problem, but at that weight she really wanted to roll, and this was the time brakes were used. Trying to get rid of that mass of energy really required brakes, and, though applied as tenderly as possible, they got hot, right up in to the red on the brake-temperature gauge which 747s conveniently have. We stopped without any blown tires, but now we had about two miles of taxiing to get back to the terminal with the hot brakes. So we very carefully and slowly taxied back, and made the entire two miles without once using brakes—no special trick, just a lot of advance planning. We didn't go fast, and I often wonder if the fast-taxiing guy really has to hurry that much. It's hard on brakes, and invites other accidents.

A Trick

Have you ever watched people park airplanes? Frequently, just as the airplane stops, it sort of lurches a bit. It's done in automobiles, too, by heavy-footed drivers.

I learned the trick of preventing that many, many years ago when I was flying copilot with a wonderful captain, Alexis Klotz. The first time out with him, he gave me one of the legs to fly. I got her up and down without much trouble, taxied to the gate, and stopped with that little lurch.

"God," Lex exclaimed, "don't you know how to stop an

airplane?" Then he went on to explain that you come up slowly to the spot where you want to stop, apply brake gently, and, just as the vehicle stops—whether airplane, automobile, or whatever—you let your foot pressure off the brakes, and there isn't any lurch at all, just a sweet smooth stop. "Watch the motorman on a trolley car," Lex said. "He swings that big lever over to stop, and just as the car is almost stopped he backs the lever off and the decelerating car stops smooth as silk." Having ridden trolleys much in my youth, it was an instant picture. Of course, there aren't many trolleys left to demonstrate with, but the idea is simple.

It's interesting how a pilot of 1938 passed part of his character and ability on to others. In the next thirty-six years of airline flying, I carried the same idea to many new copilots and taught them how to stop smoothly à la Lex Klotz. I wonder how many generations of airline pilots will stop airliners smoothly because of Lex?

I also remember a smooth-flying chief pilot of the early forties, Ray Wells, who gave copilots their final check ride to become captains. After the ride was over and the copilot now a happy new captain, Ray would give him a parting talk, which would end on a note about smoothness. His remarks were simple: "Remember, you are now the highest-paid pilot in the world—for Gawd's sake, fly like it!"

5

Habit

There are good habits and bad. In flying, habit is more bad than good. For example, most gliders that have retractable landing gears don't have a gear warning device. Surprisingly, there are few accidental gear-up landings. Many of us who fly gliders have developed some little habit to help remember to put the gear down. Personally, I have a downwind check list—a mental one—and the first thing I think about is the gear to go down. Then, again on final just before crossing the boundary, I check to see if the gear is down. It's such an ingrained thing that sometimes, when I'm flying a fixed-gear airplane, I get a sudden shock on final when I realize I haven't put the gear down. It takes a split second to realize that the gear is fixed and doesn't have to be put down.

Glider contests are generally a race over some distance depending on the lift conditions, the distance varying from 60 miles to as much as 350. At the finish of these races you dive across a timing line, make a hearty zoom, come around, and land.

A good friend of mine had developed a habit of putting the gear down right after the zoom. Well, one day, on a speed task, as these races are called, he forgot to put the gear up. He flew the entire task gear down, then dove across the finish line, pulled up, and retracted the gear! His habit pattern had told him that he was supposed to do something to the gear in the zoom, so he did: He retracted it and approached for a belly landing. Which says that although a habit may be good, it shouldn't be without thought. Making a motion mechanically and without thought—doing something from habit as an automaton would—is bad. We want habit to jog our minds, but then our minds to think!

Oh, the guy in the glider? He had radio, and as he approached on final we yelled to him that his gear was up. It registered, and he just had time to flip it down and land okay. Glider gears aren't electrical or hydraulic, with time delays, as all that stuff works. Rather, they are direct mechanical, and a shove or pull with your hands has the gear down or up. A second or two is all it takes.

Habit, with Care

I like habits such as always checking the identification of a radio aid after you've tuned it; checking the cockpit periodically from one side to the other; always making a preflight; using the check list, and many others. Perhaps you cannot really call them habits—maybe they are routines—but either way they require thinking. The habit is a trigger to start the action.

The typical chance of a habit going wrong is the check list. Habit says to use it, but it's important to read each item and then look at it, absorb it, and think carefully if it's where it should be. Our habit is to check a station's identification after we've tuned it. But does the habit simply mean flipping the audio switch, hearing some dits and dahs, but half thinking of something else, not really identifying them? Or do we make certain the dits and dahs really are . . . . — — — — (Hampton) or whatever?

Because there are many things to do in flying, and many possibilities for having one's attention diverted, it's easy to slip or slop over something without giving it real attention. We have moments of concentration on one item, like getting the . . . . — — — — of an identification, but we cannot dwell on that without realizing that it's necessary to get on to something else. It's the trick of scanning, which we've said before, and which we'll say again because it's so important. It means covering all the bases, registering what each item is clearly and then quickly getting on to the next thing. It's something like reading. How many of us have read a book and then found we're going over the words but they don't really register because we're thinking of something else? The expert reader gets what he's reading, and it registers. Well, a pilot has to be an expert reader, one in whom the subject sinks in. A pilot has to be a fast reader, because there's a lot to read.

Bad Habits

Bad habits can sneak into our flying without our noticing them. We begin to sit on the end of a runway a little too long before taking off, giving enough time for an airplane to get into the pattern and be landing before we finally pour on the coal and take off in front of him; we put the flaps down just a little above the maximum white-line allowed speed; we take our hand off the throttle during takeoff when we should keep it on so it cannot slide back and reduce power during the run. I could fill the book with bad habits that sneak into our flying.

Alcohol

I have no quarrel with controlled drinking, but any sensible person knows that alcohol and flying, driving, and a lot of other things don't go together.

The airline rules are simple: No drinking for twenty-four hours before a flight, and no drinking while on duty. This means that if I fly a six-day trip to Bombay, it's illegal to drink anything alcoholic during seven days (six-day trip plus one day before). An airline pilot is always on call, and even if he's scheduled to sit in Bombay for two days, he might be called out anytime on some schedule irregularity, and he has to be sober and ready.

Other flying isn't exactly like airline work, but it isn't a bad idea to review the possibilities of being required to fly unexpectedly, and judge drinking according to that.

How much can one drink and fly? I don't know. But I do know that there isn't any possible harm in *not* drinking, so why do it? Drinking affects some more than others, and it affects all of us more than we think. Sometimes when we feel perfectly in command of our actions and reflexes, we aren't.

I personally learned my limits while playing golf. One hot summer day we stopped after the first nine and had a beer. I had one little ole beer. Then we teed off on the tenth hole. I missed the ball completely. That's something I just don't do. It really impressed me. I wondered—not about golf, but about flying. If I had one beer, took off, and lost an engine, how good would I be? I've been very strict about booze and flying ever since.

Drinking has other ramifications, too. I talked about why some pilots lose ability, proficiency, or whatever, and said the biggest reason was loss of enthusiasm. Another big reason is drinking. Not necessarily on-flight, but off-duty drinking that puts one on the list as being an alcoholic, or nearly one. It happens on airlines, in general aviation, in corporate aviation, and everywhere else humans work. In flying, however, you cannot "get through" the day and hide it, as people sometimes do in other walks of life.

Being alcoholic brings on the shakes and makes coordination poor, but most of all it creates lack of enthusiasm and interest in the job. One's mind is too consumed with drinking to give a hoot about flying and how to do it better, or even as well.

Airlines have cases, but they pick them up pretty fast. The pilot goes downhill quickly, and it shows up on his instrument and route checks even if he hasn't had a drink within the legal period.

Of course, eventually he'll break the rules and drink while flying. In the case of an airline, a pilot who shows up half snockered is soon picked up and turned in, even by other crew members.

I guess the first time a pilot feels he needs a snort in the morning to get braced and going, he ought to stop flying and go for help.

Checks . . . and Checks

The FAA requires a check every other year for all pilots except airline, which is a good idea even if I don't like any compulsory act that means more bureaucracy. But getting a check on your own once a year is a good idea, and it shouldn't be just a hop around the field, but rather what airline pilots call a line check, made while going somewhere —because it's important to see what the pilot does during his total conduct of a flight: gathering weather, preflight, navigation, decisions en route, coming into a different airport, and all that goes with it. The airlines give essentially two instrument checks a year—one in a simulator and the other in an airplane—but they are local, how-you-fly-it checks. In addition, once a year a check pilot rides the jump seat and observes how you manage a flight, including your crew. To me this line check—and I've given and taken many of them—is the most important checking we do, because it gives a real picture of a pilot's total ability and what kind of bad habits or practices he may have allowed to affect his flying without knowing it.

Look Us Over

There are two ways to cure bad habits. The best is to take a check ride now and then with an instructor, or even a friend who's a pilot you respect and who isn't afraid to say what he thinks—and you're sure his saying what he thinks will not get to you and dull the brightness of your friendship. (That's why it's always best to get a pro.)

We can check ourselves, too, and I talk about this elsewhere. But we know ourselves best, and if we're honest and objective we can do a self-check. On some particular cross-country flight we can say to ourselves, "This is a check flight." Then we carefully look us over and analyze how we're flying to see if bad practices have sneaked in while we weren't paying attention. Looking oneself over is not all that difficult. All we need is the time and a mental attitude that we're not alibiing and trying to make ourselves look good to ourselves, but rather we're really being objectively critical.

So we use good habits to remind us to think at the proper time. We try to keep bad habits out by self-checking and periodic checks from a good pro. We all have habits; they're a big part of life. In flying we want to use the good ones and get rid of the bad.

6

Being Ahead

"Being ahead of the airplane." That's an expression we've all heard so much, we're apt to toss it off without paying attention to it. But it may be the most important expression in aviation, and certainly is worth time and talk.

Being ahead is pretty obvious: It's a matter of planning in advance so nothing jumps up and surprises us. "Planning" and "being ahead" are synonymous. But what do we think when we plan? It's broken down into two parts:

1. Advance.
2. What if.

"Advance" planning is simple. You are going to fly somewhere: How far is it, how much fuel will it take to get there, do I have the maps and equipment to do it?

"What if" planning is different. It asks what if, when up there, the weather ahead goes sour, or the head wind is twice that expected, or an engine (maybe the engine) quits?

Actually, you cannot do enough "what if" planning. But let's remember not to let "what if" planning turn us into nervous Nellies. We've got to look at "what if" with cool objectiveness, as something to prepare us for eventualities and *not* something to worry us. It's one of the parts of flying that cannot be ruled by emotion, because if "what if" thinking gets emotional, you'll never get off the ground—you'll be too scared! Neither will you ride in an automobile, or even get out of bed!

But objective, unemotional "what if" planning is a flying necessity. Being ahead applies to all flights. It isn't reserved for long trips. Even a circuit of the field calls for thinking ahead. It should be part of a pilot's character; in a good pilot it's automatic.

Let's look at a short flight from our home airport to one 25 miles away, where we'll go to visit a friend and howdy about airplanes. What's 25 miles—a flip around the corner. But it's surprising how much planning and thinking we do for that little flight.

When I was a copilot flying DC-2s on the San Francisco–Albuquerque route, we had a short leg from Oakland, California, to San Francisco: 11 miles. In DC-2s the copilot (me) pumped the landing gear up and down with a hand pump. It was hard work, and took about sixty-nine strokes. I pumped the flaps up and down, too. That route had six legs, so I was tired on the return to San Francisco, after all those takeoffs and landings.

"Say," I said to Ted Hereford, my captain, "it'd sure be nice to leave that gear down on this little hop to San Francisco."

"Yeah, it'd be nice so you wouldn't have to pump, but you're gonna pump anyway. We can lose an engine on this short flight just as fast and as bad as on a long one."

That closed the conversation, and I pumped. He was right, of course, because the airplane would not fly for sour apples with an engine out and the gear down, and neither will today's piston twins. How far you're going doesn't have anything to do with it except that, on a longer flight, being heavier, the problem will be worse. Either case, it's downhill with that engine out and the gear sticking down!

Weather

The 25-mile flight we're about to take starts when we wake up and look out the window . . . how's the weather? Everything we do in flying begins with the weather. A pilot's character makes him weather-conscious every moment of his being. As I said before, a good pilot is animallike in his consciousness of weather.

Today looks good—beautiful day, no sweat. But are we certain? Where's the wind? Humm, it's light easterly. East winds mean coming bad weather in most of the northern hemisphere. Better check what's going to happen later.

Weather looking isn't just what's en route, it's the airport we're aimed for, too. Obviously we want to know if there'll be enough ceiling to land, ice in descent, thunderstorms in the area, and all that. But even if it's a clear day, we cannot simply say, "Great!" and shove off without thinking what it's going to be like at our destination.

It may be clear, but how strong are the winds? Will they be down the runway or across it? Is it hot? Will temperature and wind bother our takeoff when we want to come home, if the field is small? Is there snow on the ground? Is the runway clear? Is it so cold there may be a chance we can't get the engine started when we want to leave?

In piston days, flying the North Atlantic I often discarded good weather alternates because of low temperatures. You'd come from Europe in a DC-4 or Connie, headed for a fuel stop at Gander, Newfoundland; alternate, Goose Bay, Labrador. But sometimes the forecast would call for Goose to have a temperature 40° below zero or worse. I could visualize my flight engineer up on a slippery wing checking fuel at that temperature, or trying to start balky engines. So I'd leave Goose out of my planning and look for a warmer alternate.

For weather the transcribed broadcast is listened to. Forecast good all day, but high clouds later, means overrunning warm air and, eventually, lowering weather. But we can watch it. So the weather is fixed and okay for the day, with an occasional glance to be sure.

It's Never for Sure

Now off to the airport—but do we have a map for the area in the airplane? And radio aid charts? What do we need them for? We know every inch of the way. But it's amazing how, if visibility lowers, familiar country can suddenly become something we've never seen before, or at least look that way.

I often fly from Van Sants, in eastern Pennsylvania, to our Philadelphia Glider Council field, 11 miles to the southwest. I know it as well as you can know any terrain. I've flown it countless times in airplane and glider. But I've sneaked between the two places in rain with poor visibility and hardly known where I was. I followed the roads, turn by turn, and set up the 260° radial of Solberg, which, with an 18-mile DME (Distance Measuring Equipment), is a fix for Van Sants, to be sure I wouldn't miss it.

We never know a route so well as not to need a map and aid information in all conditions. I got lost making a local hop on a CAVU day at Wichita and, embarrassingly, didn't have a map or radio aid chart in the airplane!

I was picking up a new Cessna Skylane. My son and I made a local hop to feel out the airplane and calm our excitement over it.

At Cessna's airport in Wichita, you take off and fly below 300 feet until out from under the Air Force's McConnell Field. So off we went, turned east, and flew away from the local airports. When it came time to go back, Wichita was out of sight—and so, of course, was Cessna's field. Not a map in the airplane. We flew west for a bit, hoping Wichita would show up; it didn't. Finally I swallowed my pride and, luckily remembering the tower frequency, finally got approach control, told 'em our plight, and squawked the transponder, and they gave us a steer. We were not very far away, either—no more than 20 miles.

"Where you been?" Dwane Wallace, who was waiting for us, asked.

"I hate to tell you, Dwane," I answered, but I did, and we still have good chuckles over that one.

But it was stupid on more than one count: no charts in the airplane, not paying attention to headings and time as we scatterbrained about, interested only in how the airplane felt and flew. A simple little laughable thing, but it bothered me enough so that I had a good talk with myself. It made an important point with my son, too, and taught him a good lesson as he watched his old man louse up by taking things a little too easy-breezy and not planning ahead.

The Ground

Terrain is part of planning. What sort of terrain are we flying over? Our 25-mile flight might be across a lake—do we want to fly over or around, where there'll be land all the way? The airport could be on the other side of a high, rough mountain—go over or around? We might want to consider using FAA's Lake, Island, Swamp and Mountain Reporting Service, in which a constant ten-minute check with FSS (Flight Service Stations) assures we'll be missed quickly and looked for if something goes wrong.

And what are the highest altitudes to know en route, be they TV towers or mountains? A little study in advance is lots better than a frantic search over a chart when we're in the air, with decreasing visibility and worry about how much altitude is necessary to be safe!

Do we cross a big city? Maybe a TCA (Terminal Control Area)? They both require planning in advance.

Terrain, of course, relates to weather. If it's coastal we

think about fog. Over mountains we think about rain and low clouds on the upwind side. The lee of mountains may have a rotor from wave action that will cause wild turbulence. Will we pass downwind of a city with its smog and reduced visibility?

All that can occur on a 25-mile flight. Airplanes have been lost on flights that short, and never found. So we don't take terrain or weather lightly because the flight is short. It's simple if planned for, but could be a disaster if not.

Before Flight

But back to our flight: The airplane's untied and the fuel checked. How much? Well, we don't need a lot to go 25 miles and back, but, whatever the amount, we must *be certain* it, and reserves, are really in the tanks! I like the Beech Sport, Sundowner, and Sierras that indicate the actual fuel when you pull the cap and look in a tank, as well as the cockpit gauges.

Preflight: It isn't any different from preflight for an instrument flight of 500 miles. How to do it is in the airplane manual, and we develop our own methods. One item not in the books much, but I notice periodic accidents because of it (or them), is birds' nests. Especially in spring, it's worth poking around the engine intakes and other orifices to look for our feathered friends' new homes. It's worse when an airplane's tied down outside, but it happens in hangars too.

We mostly think of preflights at the beginning of the day, before the first flight, but what about other flights? Do we go through an extensive preflight before every takeoff? After we've had our look and bull session at the 25-mile airport and are ready to start home, how much do we look it over?

Well, if there's any fuel added we should check the sumps for water and the tank caps for security. You don't see much water any more, and I notice that people get sloppy about it, but, as the corny old expression goes, it only takes once! And water does slip in now and then, even in the best supply systems.

I check sumps every morning, of course, and after a stop of any length during the day if there's been a marked temperature change toward cooling.

For quickie stops my preflight is a look at the tires for pressure as I walk toward the airplane, as well as who and what's parked behind me that I might blast. Then extremities to double-check if anyone has bumped into the plane while it was parked. I run my hands over the prop leading edges, because a nick can be picked up on landing and taxiing, especially on unpaved fields. When I climb in, the first act is to set the parking brake and then be sure to do the pre-start check list as carefully as ever.

Let's Go

Back to our 25-mile flight: Preflight for the day is done, the prop's clear, and we start her up, look around for traffic, and taxi out.

Now for run-up. What's so special about it? Well, first of all is position, which means off the runway, into the wind if it's brisk, and not over loose ground or a runway that's beginning to deteriorate. If the ground is rocky, or the runway deteriorating, it's easy to nick the prop, and a bad nick can grow into a blade failure, and a blade failure can be catastrophic. So "what if" planning means to run up only over smooth surfaces.

What if it isn't smooth? Then run it up on the run. Just moving a small amount helps. You ride the brakes a bit, let her creep a few miles an hour, and check the mags then. If the surface is really bad, you should be moving quite fast. Make it quick, and don't ride brakes enough to get them hot. Son Rob and I took off from the island of Tarawa, in the Pacific, for Guadalcanal in a Cessna 402. Tarawa's runway is coral and it's all loose, really bad. So we actually checked mags early in the takeoff run. Even that didn't do it, and he had to dress the props for numerous small nicks when we got to Guadalcanal.

If the surface is bad and my engines are basically good, I just don't run up, because the damage risk is worse than the good I can get from a run-up. If oil pressure and temperatures are okay, the generator generating, and the engine smooth as it comes up to power, why worry about the run-up? Too much running is questionable anyway. You see many people beating the devil out of an engine, running it for long periods at high power on the ground, when cooling is poorest, and exercising the propeller until it's about worn out. A sensitive judgment is needed to know where, how, and if to run up—which reflects on pilot character and being gentle and kind to airplanes.

As difficult as it is to believe, I see people run up in position, take their time doing it, and then pour on the coal and take off, on towerless fields, without a look back to see if someone is about to land on top of them! It's unbelievable that people are dumb enough to do this, but they are—I saw an FAA inspector do it—and periodic collisions occur because of it. It's almost too basic and intelligence-insulting to say it, but I will: For gawsh sakes, make a turn and take a sweeping look at the sky on the downwind and the approach path of the airport for landing traffic before you pull out on the runway and start the takeoff!

So run it up, check everything, and now, before the throttles open, what's off the end of the runway? How long is the runway? How far down the runway will it be safe to land ahead? After that, if she quits, which way will I go? (To that field to the right, if I'm in position; otherwise, straight ahead and pick a path through the apple orchard.) Do you know, considering how rough and strong the wind is, how high you'd have to be to make a safe turn back? (It depends on the airplane and pilot, but it's higher than you think.)

Downwind Turn

Suppose you tried to turn back: How should you fly it? Carefully! But more than that:

I don't want to open the old downwind-turn argument, but it's a fact to say that a low-altitude downwind turn is dangerous. Why? Well, gustiness will make for airspeed fluctuations and, if the airplane is descending, going through different wind velocities will cause serious airspeed variations.

But the big item is an illusion: Looking at the ground in a downwind turn will give the illusion you're going like mad. The tendency will be to feel that you have lots of speed and can slow down if need be—and that, of course, spins airplanes in.

The downwind turn must be treated like all illusions and that, again and again, is by checking instruments for the actual truth. Look inside more than out; check the airspeed and altitude by the airspeed indicator and altimeter and not by what your eyes tell you. Fly the proper airspeed. If the ground gets too close and the turn isn't working out, keep up the indicated airspeed, straighten out, land crosswind, pick a soft bush, go down a row of trees, into water—whatever looks soft—but *above* stall airspeed. Chances will be hundreds of percent better than spinning in out of control because of a stall. So *look* in, and *glance* out!

To Go or Not to Go?

Planning ahead for takeoff includes aborting, and it's tricky business. For a single-engine airplane, of course, there isn't much problem—if the engine quits, you've aborted!

With multi it's different. For big airplanes, like Boeing 707s and their ilk, the record shows that many aborts end up in accidents, and it would have been better to have kept going. Jet engines are part of the reason, because they keep gaining power as you accelerate, so you gain strength as you run.

But this doesn't apply to piston engines, not by a long shot, and piston twins, which are all marginal on one engine anywhere, and nip and tuck in the takeoff regimen. One has to know very well at what speeds he can continue with a failed engine, and be firm in his mind that below that speed he's got to abort. Runway length and surrounding terrain enter into the judgment, of course, plus the big point of minimum control speed, below which you have to abort.

Simply, what I'm trying to say is that on every takeoff with multi equipment, the pilot needs to make a judgment, based on his airplane's performance and the airport, of where he'd abort and where he'd keep going.

There's no doubt that many piston twin takeoffs are made illegally. And illegal doesn't mean being wrong in the eyes of the FAA and insurance companies alone. More important, it means being wrong with one's neck! It also means the pilot is forcing his passengers, who trust him, to take a risk which they generally haven't been consulted about. The safety of passengers is the pilot's responsibility, and taking off from an airport with a load that cannot make it if an engine quits isn't demonstrating a serious approach to this important responsibility. It's cheating with someone else's life.

The manual and numbers are there for a reason, and they should be checked for any takeoff that's questionable. Checking them for all takeoffs is better, because the check will show that some takeoffs one thought safe weren't. For example, temperature pushes required takeoff distances up to eye-opening lengths.

Let's look at the numbers of a popular twin taking off 200 pounds *under gross weight*. The numbers come out of its manual. We'll use a sea-level airport and a nice summer day with a temperature of 80°F, which certainly isn't unreasonable.

To take off and get 50 feet of altitude requires 2650 feet—2150 feet just to get into the air.

But losing an engine at takeoff speed almost doubles the

distance needed to clear 50 feet. Actually it becomes 5400 feet!

Everything is zippy with all fans turning, but lose one and things are much different: Climb rate goes to the dogs, and very precise flying is required to keep in the air. If the surrounding terrain is high, you've got a very serious situation.

Now suppose, on this takeoff, that it quits at takeoff speed and you decide to stop. Well, then it takes 3800 feet total distance from start of takeoff to hot-brake, screaming halt.

It's important to consider, also, that these manual figures are created during test flights made by expert test pilots all practiced up and ready to move fast, and under ideal conditions so they can get the best figures possible for the book. Can you fly that well every day? If you can, my hat's off to you, because I can't.

But, back on the subject, it looks a lot better to think about aborting than going in a piston airplane. Any way you look at it, sober thought should be given to what load fits the airport and the action one will probably take if an engine quits.

Note that I said action one would *probably* take. That's to allow for conditions being different from just a cut-and-dried total engine failure right at takeoff. Maybe it just dropped revs, or quit shortly after takeoff, plus or minus other possibilities that require good pilot judgment manuals don't provide. In any case, knowing the numbers gives valuable information with which to make the final, big judgment.

And don't forget wind velocity. The runway numbers come down with a head wind. But along with wind there may be turbulence that makes it difficult to fly with the knife-edge precision engine-out flying requires.

We're playing with a stacked deck, but with knowledge and planning the pilot can stack that deck in his favor.

Which Way?

Now, as we get set to take off, what's the heading? Southwest. But how southwest? 210°, 240°? It's a silly feeling to drive off and then, after getting oriented, find a big turn is necessary to get there. A look at the chart, and an eyeball heading, about 210°, is often enough, but two minutes to measure accurately builds that precision which is part of good pilot character.

How High?

We're in the air: What altitude shall we level off at? Fifteen hundred feet feels good—emergency fields within range, and yet low enough to enjoy the countryside, looking intimately into people's back yards.

We level off and set power, and it's time to look above at the clouds for a size-up: some wispy cirrus, a long-ahead signal of the overrunning. Now to the ground and see what the wind is doing there; which way shall we land if we suddenly have to? And how does the drift look on our 210° heading? If there are scattered clouds, we'll watch their shadows and get an idea of what the wind is at the cloud's level. What's the temperature outside? Real cold? If so, let's check engine settings and be certain we're not using too much power. Real hot? Then our performance is sluggish, and let's watch airspeed in the heat to keep it high enough so the engine is cool. Temperatures well above or below normal demand careful checks of the power charts.

With that all done, we're in tune with and a part of the element we're flying in, and that's part of pilot character: being constantly aware of the weather, sky, and atmosphere, and their changes.

As cruise power is adjusted, it's time to do a complete cockpit check and be certain all is set properly.

Aside from navigation, checking on weather, looking for traffic, and managing the airplane, what are we watching to be ahead of things?

With a single engine, we have constant regard for where we'd land if the engine quit. This can be from a general look when up high with room to work, where all we pay attention to is the fact that we're over terrain that has open fields. Down low we may be interested in sizing up individual fields to judge how we'd get into them.

In a twin we're not so uptight about forced landings, but it's nice to know, all the time, where an airport is. We can be pretty blasé flying with the "security" of two engines. But if one suddenly quits, the apple comes up and we're on the front edge of the seat suddenly wondering, with real interest, where an airport is. It's nice to know, and we will if we've been following a map.

Then we ought to think what we'll do if a radio fails, or all the radios. A pretty sticky problem. But we can start by knowing the official procedures for radio failure. Then, for gut stuff, what compass heading we would fly to find weather where we might let down blind and expect to break out before hitting something.

All these things we think about relate to caution and that nervous Nelly attitude I mentioned in the beginning. We shouldn't relate all the "what ifs" and advance planning in case of trouble to emotional feeling, or think that flying is dangerous. No, these are just facts of life, many of them farfetched in their chances of ever happening. But calmly, coolly, and intelligently we have thought about them, and so, if the way-out odds do come our way, they aren't a surprise. We've got some plan of action even though it was buried back in our mind a long time ago.

What's a Bad Break?

In living, aside from flying, we could make a huge list of "what ifs"—and some people do, and then overdo their concern with the list. Others never think of them, but brush them off with "It wouldn't happen to me." Well, maybe these people who don't face the realities with some plan are the "poor unfortunate" ones who always seem to have bad breaks. What we need is balance between overworry and being sensibly prepared for the facts of life, on the ground and in the air. Then, perhaps, the bad breaks will not occur so often to some.

Approach

Twenty-five miles doesn't take long, so we begin our plan of arrival soon after takeoff, and a short flight doesn't mean planning isn't as important as it is for a long one.

We want to be at a chosen altitude, speed, and location to set up our landing approach. Control of speed, especially in clean airplanes, is a big part of planning.

Say I'm doing 180 with the gear up, and I want to be down to gear speed, 140, at pattern altitude. When do I start killing speed?

Perhaps this is the biggest difference between flying a 150 Cessna, Bonanza—or a 747, for that matter. It's the business of slowing the airplane to be within gear and flap speeds where you want to put them down. The clean airplane will keep speed for a long time, especially when headed a little downhill. It isn't easy to lose altitude and speed at the same time, and it is embarrassing to be on top of the airport and still not down to flap or gear speed.

An approach starts out well with the question, How high do we want to be at the airport? Say, 1500 feet. We're cruising at 8000; when do we start down? It depends on the rate of descent you like. If it's 300 feet per minute, we have 6500 feet to get rid of. That'll take 22 minutes. We may beat the time because we'll go faster in descent.

Let's talk about speed in descent. If the air is smooth, let 'er go and keep the speed up to where it tickles the yellow line on the airspeed indicator. On jets we call it the "JBB speed." Translated, that means "Just before the bell." (Airline airplanes have a bell, buzzer, or other warning to tell when they're exceeding maximum speeds.) So to make the best time, you keep the speed right up to the peg—if the air is smooth. Sometimes weather dictates other speeds. Suppose there are summertime cumulus you've been topping at 8000 feet. Well, you don't want to go screaming down through the turbulence under that cu at an airspeed near the limit. So you stay up in the smooth, cool air much longer and then descend at a higher rate but slower airspeed, closer to the destination airport. This may mean a gear-down descent to keep slow speed and high rate.

A Point about Flaps

An aside, but an important one, concerns the use of flaps to slow down in a rough-air descent or during turbulence. Don't use 'em! The structural strength for flaps extended is considerably less than for the airplane flaps-up. So your chances of structural failure go up with flaps down, even though you are obeying flap limit speeds.

Descending fast in rough air is bad because of the uncomfortable ride, but it's bad too because of the bangs and wallops on the airplane structure.

The point is that we should plan the descent. It shouldn't be a matter of vacantly flying along and then waking up to the fact that it's probably time to shove her over and get on down. That hit-or-miss method results, often, in arriving over the field too high and going like crazy, or getting low far out to wallow around in rough air for a long, wasteful time.

Sorry!

I had one of those embarrassing too-high situations on a Connie flight from New York to Shannon, Ireland. We flew at 19,000 feet. A nice night and a nice crossing. In those days we had bunks on board, so about mid-Atlantic I crawled in for a couple of hours' nap. I asked to be awakened about an hour and a half out.

I slept well and when called climbed out, washed my face, got a cup of coffee from the front galley, and stopped at the navigator's table on the way to the cockpit.

"How's she go, Ray?"

"Seems fine—fixes aren't great, but I got a good check on Jig." That's what we called weathership *Julliet* then.

When I settled in the seat we had about an hour to go, and I mentally computed we'd start down in about twenty-five minutes. I flipped up the audio switch for the marker beacon as I was checking how things were set in the cockpit. I suppose I set the marker audio in anticipation of the Killkee marker when we reached the Irish coast. But yipes! The marker was coming in loud and clear. We were almost over Shannon! And at 19,000 feet!

We'd picked up a strong tail wind. Navigation then wasn't as it is now, with inertial which tells, all the time, wind direction and velocity, your groundspeed and exact position.

I called Shannon and told them I was 19,000 feet over the field, requesting descent clearance. The controller, with a suggestion of humor in his Irish voice, cleared us and we augered down in big circles. I felt very foolish.

Of course, that was a navigation error, but we can be almost as silly and, what's more important, inefficient if we don't plan descents on an estimated basis. It follows that our navigation should be as precise as possible, keeping tabs on position, speeds, and estimates all through the flight.

Knowing position allows keeping up speed until the last moment, and that helps the air traffic system move traffic. One guy slowed too soon will string out and hold up a long line of airplanes.

We keep our speed up, aiming at some point where we'll start to slow, and this point can be an outer fix at a busy big airport, or a couple of miles from the small country airport where we're going to land.

Slowing

The key is in knowing how long it takes your airplane to get from yellow-line speed down to the first-flap or gear speed. To learn we just go out and try it. How do we slow an airplane; what's the technique?

The best way is to come down to the desired altitude, *level off*, and reduce power. Notice I emphasized "level off." If we start slowing while the nose is down, and we're descending in a clean airplane, it's going to take much longer, and we'd better allow for it. It's best to get down, level off, cut power, and wait for speed to drop. It will be quickest in that manner.

It's useful to experiment with your own airplane and see how long and far it takes. On a nice day try flying at speed just below VN_o and then hold level, cut back power, and time how long it takes to get down to gear or flap speed. This will have to be done with some power for protection against loading an engine, icing, and all that sort of thing. You cannot pull back to full idle and just sit there—except in a jet, which is another of its nice features: At full idle it doesn't load up, ice up, or quit.

But having a precise idea of what slowing up takes in time and, therefore, distance will be a big help in all approaches. It's worth trying during different descents, too, so if ATC says, "Come down and slow down," you'll have an idea how to do it and what distance and time it will take.

An annoying operation, which I've seen too often, happens when the guy pulls the nose up and climbs or zooms to lose speed. It's an old throwback to the military overhead-fighter-landing technique.

In normal operation it's hazardous because someone might be above. ATC may have another airplane close above our zooming friend. VFR, there can easily be another airplane above and slightly behind. When the low man pulls up he slows; then the higher airplane catches up, and is run into by the zooming airplane. The lower, zooming pilot couldn't see through his roof, and the higher pilot couldn't see through his floor.

But also annoying about this pullup is that it wastes energy and is stupidly inefficient. We can plan to get back some of the energy used in climb by a neat, well-done descent, but to not plan and to toss away energy in a zoom is awful!

This is all part of being ahead of the airplane, planning what's going to happen well down the road. A descent has many factors: distance, time, weather, the need to slow early, and, on some days, the chance to let 'er rip right up to the field. Putting all the factors together and deciding which plan fits you on that day, for that flight, is another of flying's fascinations. No two trips are alike.

Landing

We approach our 25-mile field. Make the cockpit check before landing. Why? We just took off, nothing is different. Oh, we did change tanks to balance that one wing; better go back to both. Yes, check and check; it never hurts, and it develops a habit pattern that will be a good part of pilot character.

Now, where is the field? We want to see it early and begin to spot traffic, begin to fit myself in. How's that approach? Let's see: big trees to one side of the approach. The wind's a little cross and will cause turbulence off those trees. Any wires? Sure? And where do we park? Is it best to plan a long run out and not hold up people behind us, or do we land short and turn off at the parking area? The location of parking areas and turnoffs affect approach planning; we want to help expedite traffic and make taxiing a smaller job for ourselves.

Base leg and final: Plan them short, get on with the job and on the ground. Unfortunately, many pilots seem to think a base leg should be miles from the airport, and final a cross-country flight in itself. If we're cutting ours to a reasonable short pattern, we have to take hard looks "downstream" to make certain some "way out there" pilot hasn't made a long final to be potential collision traffic for us. Too bad some pilots make these long finals—they're dangerous as well as inefficient.

Part of airport planning is knowing the length of the airport. It's on the maps. If it's long, we've got all kinds of room to play with, and 4000 feet, for example, when we're in a light aircraft, is really a couple of airports, so if it's useful to land long to save taxiing, we can do it safely. If the field is 1600 feet, we're more interested in getting in short.

On takeoff I like to use all the runway, even if it's 10,000 feet long. Intersection takeoffs save time, but I use them only if the obstructions and "junk" off the end of the runway are hospitable enough to accept me if an engine quits. If that area isn't hospitable, then I take all the runway.

Approaching the Bankstown airport in Sydney, Australia, in the Cessna 402, my son flying, we looked up the runway length. It was just 3000, which, while enough for a 402, isn't anything extra, and it didn't seem very long for an important airport.

"They must mean meters," I said.

"Yeah, I guess," Rob responded.

There was considerable traffic, and we were even mildly cut off on approach by a Cherokee, so attention was more on these things than the approach path. Finally we were lined up and set. Rob took a hard look, since we were a little high, and the runway suddenly looked small and tight.

"Keeripes! They do mean feet!" he yelped as he went to work and got us down to a lower, slower slot for a busy last-minute landing. When flying outside the country, it's a good idea to be sure about that meter-feet business, for both runway lengths and terrain altitudes.

Planning also means looking for a parking spot. Are there tie-downs; does it face the wind—downwind or cross? Can we get out again when we're ready to go?

We shut down and make certain the secure-cockpit check list is read, controls locked, airplane tied and chocked. Even though the flight is over, we're planning ahead for the next flight, being certain we've got a workable, undamaged airplane that's in position to get out and going.

Long Flights

Longer flights don't require much more thinking ahead than short ones do, except for weather.

Thinking ahead about weather is an endless job. You can never say, "I've got it locked in, I don't have to worry, it's in the bag." It never is.

I learned that, or was awakened to this possibility, on one of my earliest copilot flights. The captain, new and just a mite cocky, looked over things for Pittsburgh (we were at Newark), and even though the weather wasn't very good, he satisfied himself that it would be okay at Pittsburgh when we got there. "It's in the bag," he said.

We flew west, drank coffee, had a giggly conversation with the hostess, and didn't check the weather ahead. When we finally started letting down to Pittsburgh, it came as quite a shock to find it below limits with freezing drizzle. We collected our wits, plus information, and finally landed at Columbus, Ohio, with a much less cocky captain than we'd had at takeoff from Newark. I've never liked the expression "It's in the bag!" since.

For safe flight, the quest for weather information and knowledge of what it's currently doing should continue right up to approach, when we ask for a last wind check on final.

The pilot who keeps a constant picture of the weather

over the countryside is the safe pilot. There are lots of details in my book *Weather Flying*, but the one I want to point out again is that, once airborne, a good pilot keeps collecting weather information, creates a picture in his mind of what it is doing under him, ahead of him, at his destination, alternate and the direction from which weather moves in his part of the world.

All this is related to the fuel he has, and whether that will take him to a sure-fire clear-weather airport if his destination goes sour. The alternate doesn't have to be clear, but it must be within useful limits and, what's most important, positively going to stay that way or get better. An alternate's weather must be holding or on the improving side for at least twelve hours after our ETA (Estimated Time of Arrival)! If, within that twelve hours, it's forecast to deteriorate, then it isn't a good alternate!

Any forecast is only as good as its time span—the older it is, the poorer its chances of being good, and, of course, the newer, the better. Which says to pilots: Be certain to notice when the forecast was made, then put that in your mental computer as a factor in trying to judge what the weather is going to do. An old forecast should be looked at with much suspicion.

Weather information isn't exclusively a matter of destination, either. What's in between is as big a part. Where's the top, the ice, or thunderstorms? Listen: Listen to broadcasts, listen to the frequencies used by others, and pick up all the information possible. If it isn't there to be heard, then ask. What are the tops? Where's the ice? Thunderstorms? Listen and ask, and never let up.

Surface weather, aside from telling what a destination is

like, is very useful for the VFR pilot in telling him when he may get into trouble.

The ceiling reported is important for ground-airplane clearance, but visibility is the big item. If the visibility is at or near VFR minimums, the next thing to check is precipitation. The weatherman who said he could see three miles was looking out a window, or standing on a building that wasn't moving into any rain or snow at high speed. That makes a big difference.

Rain smearing a windshield cuts visibility. To prove how much, make a little test the next time you're flying in rain: Simply look ahead through the rain-covered windshield and see how far you can see. Then look out the side window that isn't getting rain head on, and note that you see much farther. It's too bad the industry cannot come up with something to improve windshield visibility in precipitation. It would save lives, especially among the VFR-flying people.

Three miles visibility reported with rain is often a half mile from the cockpit. If you plan to navigate and miss mountains, TV towers, and other terrain VFR, don't let a three-mile weather report lull you into thinking things will actually be seen three miles ahead if there's precipitation.

En route, VFR, mountains are where the trouble is. Weather stations on each side of the hills may give acceptable VFR reports, but if their ceilings are below heights in the mountain and there's precipitation, the mountains are probably in the stuff, with no way through. Sneaking through passes will be difficult and dangerous. Again and again, VFR flying in marginal weather, and especially weather with precipitation, is a dangerous business, and many times more so in mountains.

So we think ahead by getting weather ahead; we also think ahead in the sense of the clock. What time of day is it? What time will it be when we get to where we're going? Will it be near dark? If we're delayed by winds, or a late start, will it be dark? What time does it get dark this time of year? How long is twilight? It's longer in northern latitudes than in southern. In Florida and other places toward low latitudes, it gets dark right after the sun ducks below the horizon. How does the night fit our flying ability? Do we want to be out after dark?

And always, in planning, is the fact that weather gets worse as it gets dark. So we think ahead to time of day, our ability and willingness to fly in the dark. There's nothing sissy about not doing it in a single-engine airplane. Again, remember that weather is going to deteriorate after dark.

A dark, dark night in open country, with no moon and few lights, requires thinking ahead to the fact that it's the same as flying on instruments. Useful outside reference is nil, and if you cannot fly instruments, a black night isn't the place to fly.

Planning and being ahead isn't all covered in this chapter, nor all the chapters written in all the books, but I will bring up points like these again and again. Hopefully, this will awaken a desire to think more about being ahead of the airplane, to think of new ways to do the job, new things to plan for well in advance, from the longest complex flight to the shortest hop around the field.

When you're in an airplane, flying instruments, and the airspeed starts to increase during a maneuver, you say to yourself, "The airspeed is starting to increase. I'd better get to work and stop the increase, because if I don't it will get out of hand."

Now, you don't think all that consciously, but you know it inside, and that, in a micro way, is being ahead. It demonstrates that all flying, from a tiny passing action to long hours en route, needs to be guided by being ahead.

It will bring tranquility and security to flight.

Being a firm believer that teaching only teaches one how to think, I list below some things to think about in relation to being ahead:

> weather
> terrain
> power loss
> radio loss
> fuel
> traffic
> the airport area
> airplane performance
> equipment for the flight.

7

Looking

There's a lot of looking to be done in flying—outside looking, inside looking, maps and paper looking—enough so that looking has to be an organized, trained operation.

We call it "scanning," and it's as good a word as any, except that scanning often has the connotation of superficiality, glancing at something without really taking it in.

Well, this cannot be, and in scanning while flying we have to look quickly but register what we see. This is becoming more and more important as we enter the computer age, with buttons to push and lighted, digital numbers to read.

We glance at the old-fashioned "steam-gauge" type of oil pressure indicator, and if our glance shows the needle to be about where it should be, that's enough to know that the oil pressure is okay. But in digital readouts the oil pressure is given as a number, and you have to read the number and let its value sink in.

I had this come strongly to me with the first 747s and the new Inertial Navigation System (INS). It's a simple system so long as you program it correctly: that is, push the right buttons when setting it up.

A New Era

My rude awakening came on a flight from Paris to New York. The procedure was to program the INS for the last point of land and then 10° latitude increments after that. The automatic pilot follows the INS, and all you have to do is sit back and keep from being bored to death.

This flight's last point of land was Quimper, a place out on the west coast of France, near Brest. From there on it's the Atlantic until Nantucket Island, off the coast of Rhode Island.

The first check over the ocean after Quimper was 48° north latitude and 8° west longitude. That's the boundary between French air space and Oceanic Control.

Well, we passed over Quimper, and the autopilot made the airplane turn toward 8° west. The turn seemed pretty big, so I started to check things, and the first thing I noticed was that the INS said it was 427 miles to 8° west. I knew that wasn't correct, because it was about 157 miles. Something was wrong, and it made me sit up and start looking for the problem, fast!

My copilot and flight engineer were in on it by this time, because they'd noticed the big turn too.

"We've probably put in the wrong numbers, so let's check that first." We did, and all three of us saw it was 4's, o's and 8's. That wasn't the problem.

Now it felt more serious, because something was wrong that wasn't obvious. "Maybe we put the wrong position in at Paris when we erected it."

So we checked that. It was okay.

Of course, all this took only a few minutes, but it was perplexing and worrisome.

"Let's start all over," I suggested. So we went back to the Way Point selector to take another look at 48° north and 8° west. And there it was—we'd put 40° rather than 48°, but our quick first glance as we flipped the selector switch to read the numbers had shown 4's, o's, and 8's, and so it had seemed okay. We hadn't actually stopped to read the numbers carefully. We'd glanced at them in the old "steam-gauge" fashion pilots have been accustomed to for so many years, and taken the glance for granted.

We corrected it and went on our way in a few minutes. No problem.

But I realized then that we were in a new era, that new times and methods had entered the world of flying. The computer age was with us, and glancing wasn't going to do the job any more.

Self-Teaching

Starting a self-discipline program to scan but *see* showed me that I hadn't been looking and registering as well as I should have all along. Yes, you glance at that oil pressure gauge needle and it's about where it should be, so it's okay. But by reading more precisely you might notice that the needle isn't exactly where it was. At first glance it seems so, but on closer inspection it's really riding just a hair lower than it has been. Why? So you check other things, and perhaps pick up the beginning of a malfunction before it gets serious.

Teaching ourselves to look quickly and yet see opens new worlds in many ways. We read better and see more, because we don't look back in a "how's that again?" double-take way so much. And if we don't take that second look, we have more time for looking.

But how do we train ourselves? One simple start is getting a digital watch so that whenever we check the time it's necessary to read the numbers rather than just glance at hand positions, as in older watches.

The flood of electronic calculators means that one can be purchased inexpensively and used for daily math chores, balancing checkbooks and budgets, and so on. It's excellent practice, and the face of one of the calculators isn't much different from an INS keyboard. This isn't to train people who'll never have one to use an INS, but rather to learn to look and see correctly. The object is to do it quickly and accurately. Practicing with an electronic calculator at home, we'll not make a mistake and get lost later . . . of course, we might mess up our checkbooks and have a few overdrafts!

A game I like goes on while driving my automobile. I glance at another auto's license plate, then look away and try to visualize the number. I look back to check if I got it all, and correctly. The object is to make as short a glance as possible and still catch the numbers correctly. It's an easy source of practice, and helps relieve driving boredom. Of course, it's a good idea to scan the traffic too!

This all fits in with the fact that we cannot stare and leave our attention too long on one thing. For example, let's try to find that intersection ATC just told us to proceed to. We've got to hunt it out on the chart. This would be easy on a desk at home. But our desk is in the air and moving, so we can't stare at that paper too long. The trick is to glance at an area of the chart and look back to the horizon, artificial or real. While we're looking back at the horizon, our mind's memory is searching the area of the chart we looked at, like visualizing those automobile license plates, for the intersection. Then another glance at the paper again, back to the horizon and a flick look at airspeed—but basically if we reference the horizon our heading and airspeed will be within bounds. All this is an excellent argument for an autopilot; even the simplest single-axis is a big help. Also, if I cannot find that fix easily, I'm not embarrassed to ask ATC where the devil it is! But that doesn't relieve me of learning the art of scanning.

Two keys we're trying to get across are: we can't stare, and when we look back the best thing to look at first is the horizon.

Outside Looks

Airplanes and the air traffic system we fly have a lot to look at and much not to overlook. It's no place for daydreaming.

Flying VFR, we are interested in who's outside, and to see him (or them) we set up a roving "look" pattern. AOPA (Aircraft Owners and Pilots Association) has devised good training for this. It's especially important in climbs and descents, because then controlled and uncontrolled traffic have their greatest "mix." Climbing and descending are serious and important times for looking out, controlled or not. Most airlines have a rule that no paperwork is done in the cockpit during climbs and descents.

But we need a roving look inside, too, a way to check all our knobs, switches, instruments, and gadgets. A good way is to systematically look from the left side of the cockpit area to the right side in a sweeping, but careful, glance.

In my Cessna Skylane I check it something like this: left side door latched, marker beacon switch location, primer in, master switch okay, vacuum, airspeed, gyro vs. compass readings, heading, altitude, omni frequencies and radials set as desired, how radios are for use (what receiver and transmitter are switched to come through the speaker or out through the mike), DME, cowl flaps closed if in cruise, wing flaps up, fuel-tank selector, the electrical switches to check what's on and off, the carb heat, propeller, instruments, engine instruments, fuel gauges, mixture, and right door latched.

All this is simply organized by a left-to-right coverage of the cockpit. Left to right, up and down en route, to be certain it's all covered.

Obviously, I don't look at all this every few minutes of flight, but I do look periodically: before takeoff (in addition to the important written check lists), when airborne and settled in climb after a good look outside, when leveled off, periodically during cruise, and any time the flight situation has changed, like going to a new altitude.

Naturally, there are things we look at very often, especially when on instruments, and I call that the "constant" scan. That's covered later.

The first scan we've been talking about I call the "big" scan, and it's necessary even though some things would seem to be set for good once you've checked them.

But a passenger moving to reach back and get the thermos could unlock the door; a pushed button might have you transmitting on #1 transmitter when you thought you were using #2. You may have climbed and opened the cowl flaps, and because a radio call distracted you just as you leveled off, the cowl flaps never did get closed.

Things have a way of changing, almost mysteriously at times; some are overlooked or forgotten. The systematic, periodic big scan takes care of the problem.

A good pilot's eyes never sit still and stare. I was educated to this by the great pilot of early racing fame, Al Williams. We were with a group of people, and one man had a beautiful German shepherd dog with him. We admired the dog, which, with roving bright eyes, looked and "smiled" at us all.

"Look at his eyes," Al said. "They never stop. He'd make a great pilot. That's the way a pilot's eyes should move, covering everything often."

I was about eighteen then, with some 200 hours. I'd never thought about moving eyes and looking around, but Al's remarks put it firmly in my flying mind. It was a good lesson.

The constant scan relates to instrument flying. This has

two facets: One is looking often, and the other is the priority of what we look at.

We want the airplane to stay in position. This relates to heading and pitch, which relate to the artificial horizon. If the wings are level and the nose near the horizon, the airplane isn't wandering very much. So the number one priority instrument is the artificial horizon. Our scanning starts there, does the rounds of other instruments, and comes back to the artificial horizon. We start and stop turns with it, originate and stop climbs and descents. It's number one!

If we're flying an ILS, OMNI, ADF, or whatever, we aren't flying the needles, we're flying heading in order to put the needles where we want them. So to fly heading we fly, again, the artificial horizon first, directional gyro next, and then check to see how we're doing with the "location"—that is, the ILS needles, OMNI needle, or ADF pointer.

To repeat and sum up, because it's important: If the airplane's relationship to the horizon is constantly monitored, the airplane isn't going to be far from where you want it.

The other instruments, airspeed, altitude, and so on are important in the problem we're doing and must be covered frequently. And, to repeat because it's worth it: Check that the proper frequencies are set up, and the radials in OMNI windows. You'd feel silly going out the 90° when it should have been the 60°!

So it all comes back to the point that to be a good pilot one has to be able to scan quickly, often, and accurately. This may sound as though we develop a sort of frantic eyeball race to cover all bases. Actually it isn't; the mind moves fast, and has the capacity to handle the problem with time to spare if it's used properly. Mostly that means relaxed. The mind moves fastest and best when relaxed. What makes a great athlete? One thing is the ability to go full out while still free and relaxed. Our goal in practice is to scan quickly and easily in a relaxed fashion.

I remember giving an instrument check in a Constellation to Tex Butler, one of TWA's senior pilots, and a good one. I had Tex coming down the range leg at Kansas City for a low approach with one engine feathered and the control power boost off, a handful by any standards. He was doing a beautiful job, and eating an apple at the same time!

So we want to scan, cover all bases frequently, and be able to eat that apple too.

8

Listen

We do two things with radios: speak and listen. There's a lot to be said for just listening. In flight we learn much without ever using the speaking part of the equipment.

After engine start, turn on the radio and hear what's going on around the airport. If there's an Automatic Terminal Information Service, we get the active runway, altimeter setting, wind and weather. But without an ATIS we can gather much useful information, even on a little-used airport. Listen to unicom and pick up the active runway, get an idea of the traffic, and start to feel yourself part of the airport and its action.

Listening to the tower at a busy airport that doesn't have ATIS will reveal the altimeter, wind, and runway as he gives it to others. We can also discover what sort of taxi route the tower is demanding, and if the mumbo jumbo of the taxi route and taxi-strip letters is baffling we can look them up or watch others go out before it's our turn, so that when we finally go we can be very pro-like, say "Roger," and taxi without getting lost or asking the tower for directional guidance.

Despite all the procedural language printed and preached, each tower does things a little differently. The physical make-up of the area changes procedures, and listening will pick this up—like the tower giving all VFR traffic a right turn instead of left after takeoff, or some other type of flight path.

Before the runway we get an idea of the traffic ahead and behind and where we'll fit in. If it's an instrument departure, listening to another pilot's clearance gives a clue to what's coming for us when the man calls and says, "I have your clearance." Hearing in advance gives a chance to get the right charts organized, radios set, and the radio repeat-back easy to do.

In the air, en route, all kinds of useful information are available. If we're on instruments, we listen, of course, to ATC and especially for his call to us, but the calls he makes to others tell us much: who's above, below, ahead, and behind. The airplane ahead is a preview of what's going to happen to us. The controller tells him to change frequency; we hear the new frequency, and write it down or set it up on our other transmitter. Then, when the controller gives us that same frequency, we're ready to answer crisply and change frequency without floundering.

Approach Info

Approaching the terminal airport, the airplane ahead is a look at what type clearance is being given for approach. Hearing that, we can be all set, with charts, radio, and our minds, to do it simply.

Now and then, of course, we get something different, and that's part of the game, but the big percentage says the pilot ahead will do what you're going to do.

The pilot underneath is a good indicator of when you'll be let down, especially in a holding pattern, because when they clear him to descend, you're next.

And while this instrument flight goes on, we pick up altimeter settings, weather at the destination, and an idea of how things are going down there by the regularity of approaches being completed. If they move along fast, then the weather is as good as they say, but if approaches are slow, or someone misses one, it's time to think that you may have to go to an alternate.

It's Exciting

Bringing an airliner into a busy area like New York is an exciting and interesting spectacle. I've come to JFK from Europe on busy afternoons with marginal weather. The sky is electric with radio chatter, pilots being given clearance, asking information, changing flight routes, and a hundred different things. Listening to the action, the constant chatter between airplanes, ATC, and, on another receiver, the company messages, we have a front-row seat to a constantly changing drama. It's a great lesson in professional action. We hear the crisp acknowledgments of the good pilots, the cool pilots. And we hear the excited voice of a nervous pilot who isn't thinking as well as he should be and is allowing emotion to get in his way. It's funny, yet sometimes sad, and once in a while scary, to hear a guy asking for weather in a tense, high voice just a moment after approach control has broadcast it to all aircraft.

On the North Atlantic, weather is broadcast on high frequency at various periods during the hour for major coastal terminals. I've heard a pilot make a frantic request for weather when, at the same time on another frequency, the weather was being broadcast! And I'm certain he had equipment to hear it!

Light Airplanes Too

Flying a small airplane cross-country has the same listening opportunities, and there's a world of information available by listening to frequencies on which others are talking.

There are weather broadcasts, although in some areas they are poorly done. But listening at the advertised times often gets weather without asking for it. The Flight Service Station frequencies can bring a wealth of information as others call for weather and information.

In some parts of the U.S.A. there are Flight Watch stations, which deal only in weather. They give weather on a demand basis and also welcome weather information from pilots which is an aid in forecasting and informing other pilots. There's big chatter back and forth with Flight Watch stations. The chatter is so extensive that we rarely ever have to make a call ourselves, because a little patient listening brings all the weather we need.

It's obvious, when listening on their frequency, that a lot of people haven't learned the art of listening, because you'll hear the same weather being requested again and again, perhaps one moment right after another. It's obvious that there are a lot of trigger-happy, grab-the-mike minds in flying. This is the mind that's suddenly set off by an idea such as "I wonder what the Oakland weather is." And before the thought process goes any further, his hand shoots out for the mike and he's asking for it. How nice if, when he wondered what Oakland weather was, he decided, "I'll listen for a minute and see if I hear it." By doing that he frees air time for the pilot who may need something badly. Someday our Oakland pilot may be in that spot but will not be able to get a word in because another guy just like him is using the air time. And it'll serve him right!

Where's the Weather?

Listening for weather is done in two ways. The first is to monitor the scheduled broadcasts from omnis serving FSS at :15 after the hour, and Transcribed Weather Broadcasts, which are transmitted continuously from radio beacons in the 200-to-405 kHz range. You get it on the ADF. There are a few continuous broadcasts from omnis, too.

Unfortunately, at many locations these broadcasts are done on a slipshod basis. They hold a low priority on the FSS list of duties and are done only after everything else is out of the way. Many times, therefore, the broadcast is old and incomplete, with important parts skipped. It's spoken too fast by the harried operator, so you don't know what he's talking about.

It's a chicken-and-egg, dog-chasing-tail sort of thing. The broadcasts aren't done well because the FSS man has too many other things to do. A lot of these other things are answering calls from pilots who want weather information! So if fewer people asked for weather, the FSS man would have more time to do a better job of making up the broadcast, more people would listen, and he'd have fewer phone and radio calls. But to get people to make fewer calls the broadcasts would have to be better first . . . which can happen only if the FAA gives them a higher priority. So whose move is it? Crazy!

Many times, where I live in Vermont, I listen to the Burlington transcribed broadcast, which has to be one of the worst in the country. I get everything from nothing to a rare fair show. The result is that most times I have to call on the telephone and use up the FSS man's time, even though I try to make the call short. It would be very useful, very beneficial, and save money if the FAA provided a crisp, accurate, and up-to-date continuous weather-information service.

But despite this outburst of mine, about a pet peeve, it still pays to listen to broadcasts for weather. There are places that do a good job. Wichita, Kansas, comes to mind, and I've never tuned them in without getting most all I wanted in a well-done fashion.

For those who have HF radios, it's worth knowing about the East and West Coast hourly broadcasts of weather for various key terminals. These are used mostly by the trans-oceanic flights, but there's no reason why somebody coming into Boston or San Francisco, say, cannot use them too. On good reception days you'll also get Canadian stations. The frequencies for the East Coast are: 3001, 5652, 8868, and 13272. For the West Coast they are: 2980, 5519, 8903, 13344. The best reception will be on the high frequencies during the day and the lower ones at night; the old chestnut says, for a guide, "high noon, high frequency."

Besides weather, the omnis send Airmets and Sigmets when appropriate. Appropriate means when the weather service thinks some kind of disastrous weather may seriously affect aircraft.

So clever listening includes standing by on the omni station nearest you, because you'll pick up these "Sig-Airmet" broadcasts and hear the omni giving weather to pilots who have asked for it, and it might be just the weather you want. There are a lot of interesting things to hear from an omni: weather, Sig- or Airmets, information broadcast to other airplanes, and pilot reports.

Even on Instruments

The frequencies pilots call FSS are interesting listening too. Even if I'm on an ATC frequency during instrument flight, I try to monitor either 123.6, 122.6, or 122.1 MHz. Weather information pours in on these as people ask for it.

Of course, where it's installed, the Flight Watch, currently on 122.0, is the weather frequency to monitor.

Or VFR

I often fly from Van Sants airport to Montpelier. If possible I go VFR, because the route ATC clears you to fly on instruments adds about 30% to the distance and takes you over bad terrain I don't like. The listening part of that flight, VFR, goes something like this:

While warming up I tune 374 kHz on the ADF and catch the Newark transcribed broadcast for the latest weather. I set up 122.8 to hear if anyone is coming into Van Sants and what wind and runway Mary Smela, who generally handles the unicom, is broadcasting.

In the air it's about a 031° heading, and tune the Solberg omni to listen what they are giving from the New York area to requesting pilots. I try 122.6 and set it low-key, but enough to hear. So we've got 122.6 and the omni giving all kinds of stuff. Sometimes I alternate 122.6 with 123.6, which seems to be more local. 122.6 sometimes is too busy, as you hear jets 300 miles away revising their flight plans or asking for weather; 123.6 isn't so widespread.

Farther north I tune Hugenot omni to see what his action is, and as I pass Wurtsboro, N.Y., I try 123.3, because it's the glider frequency, and I want to hear if any of my friends are aloft. This frequency gets cluttered at times because of flight schools.

North of Wurtsboro I tune Albany omni, and they are

pretty reliable about their :15-after-the-hour weather broadcast. One day the man giving the broadcast ended it by
saying, "Anyone reading this broadcast please call Albany on
122.1."

I got on the horn and called him. "What'll you have?"

"I'm checking to see if anyone is listening," he replied.

"You bet I am, and I appreciate the good service. Thanks
very much." I could almost feel him getting happy over the
idea that he counted to someone. If we all listened, and now
and then dropped a note or called at an unbusy time to
thank these people, we might get better service.

Albany is a busy 123.6 station, and its omni gets a lot of
action too. I never go by there without hearing most of the
weather I want.

North of Albany it's over to Glens Falls omni. That station
seems to get lots of action, and they talk a lot too. So anything I didn't get going by Albany, I'm sure to hear from
Glens Falls.

A little north of there I can read the Burlington transcribed broadcast on 323 kHz, but, as I've noted, this is
sketchy. Sometimes I get good dope, but most times I don't.

The last lap is the Montpelier omni and 123.6. Listening
to that for 40 miles, I learn the altimeter setting, wind, runway in use, and how many 150s or Cherokees are shooting
touch-and-gos. At the Williams beacon, 6 miles south of the
field, I make my first radio call and tell Montpelier, on 123.6,
that I'm in the area and have the information. I call on final
for the traffic's benefit—downwind if it's appropriate, and
after I've landed, just to let 'em know I'm on the ground and
out of the way. Total radio use time for the flight: a minute.

There's a lot of information up in the ether. It's there for all of us to hear. The mike is something that hangs on a hook, or is a button on the stick, to be used as little as possible and, when it is, in proper fashion—which I'll talk about next.

9

Talk

     I'm an exponent of talking as little as possible on the radio.

That old thing about the parrot, which is a bird that says a lot but doesn't fly very well, while the eagle . . . well, you never hear him speak.

And things about keeping one's mouth closed so as not to put one's foot in it.

These seem to apply in flying, along with other points that make too much conversation on the air a bad thing. For one, it uses time which may cut out the chance for someone who has something important to say. Perhaps that's the most awful result of all.

Putting your foot in your mouth is a valid point, because by talking too much you will finally say something you shouldn't, or admit to doing something improper, and that is bad in this world where Big Brother is always around and most towers, ATC, etc., have tapes which record all conversation. This shouldn't scare or intimidate us, but it should make us give a little thought to the legality of what we're about to say.

Of course, if it's an emergency, or something approaching one, then we shouldn't be too timid to say, for example, "I'm not sure where I am—how about a radar fix or DF bearing?" The heck with legality then. We can argue that out, after a safe landing, in the warmth of an office and, if needed, with the services of your sharp lawyer.

Another sorry thing about too much radio talk is people who allow themselves to get emotional and irrational. This generally comes from arguments between aircraft, ATC, towers, etc. Arguments like that belong on the ground, over a telephone or face to face.

Who's Boss?

Which brings up the important point about asking and telling. I can demonstrate what I mean by a little incident which occurred when I was letting a new copilot fly. He did a nice job of takeoff and climb and the general chores. There were thunderstorms. It was daytime and we could see them and they were easy to dodge. He grabbed the mike and said to ATC, "There's a thunderstorm ahead, and I want permission to turn thirty degrees to the right."

"Son," I said, "you just made your first mistake."

"What did I do?"

"You asked 'em when you're supposed to tell 'em. You should have said, 'There's a thunderstorm ahead and I'm turning thirty degrees right to avoid it.' "

Now, this may sound like an old-fashioned, individualistic pilot trying to keep a tight hold on his individualism, and that's what it is, because without it we're in deep trouble. But that isn't the only reason.

Ducking a thunderstorm is related in a way to an emergency: It isn't one, but if you don't miss it and plow through instead, there may well be one. So in this case you don't ask 'em, you tell 'em. Now, if it's a matter of wanting an altitude change to stay on top of a choppy stratocu deck, or fly where the winds are better, then you ask 'em. If the stratocu deck has ice, you ask 'em once, and then, if they don't get you a change and the ice is getting worse fast, you tell 'em you're going off course and changing altitude.

Unless it's a dire emergency, you listen a bit, after you tell 'em, to see what the reaction is. They might have a more serious situation than yours, and it's worth listening to learn what it might be before you act. But you've established who's who and what's what by telling instead of asking.

Being more aware of telling than asking means that the pilot is still in command of the airplane. More and more people on the ground get into the cockpit and try to fly your airplane. ATC says turn this way, turn that way, go up, go down, slow down, speed up. I've had them argue with me that there wasn't a thunderstorm in front of me when I could see it big as life and ready to engulf me if I didn't move over. Towers tell you to land short, land long, taxi slower, taxi faster. The FAA spells out how to move the stick and rudder and throttle, and tells you what to use to go up or down, when any good pilot knows that different conditions require different techniques and sometimes combinations of two or more. Ground people like to fly your airplane. ATC would love to be able, by telemetering, as is done with space shots, to do the heading changes of your airplane from the ground and not have to bother having you do it. You'd just ride along! Think I'm kidding? I'm not; it's been talked about in Washington circles. The one time, however, that all these people disappear is when something happens. At the official investigation you and you alone were flying that airplane, and you take the blame. The ground people mysteriously didn't have a thing to do with it! Well, if you're going to take the blame, you want to be the person responsible for the action.

Another point, of course, is that the man on the ground doesn't have the entire picture: He doesn't know the weather aloft, he doesn't know how the airplane is acting, he doesn't know what the fuel problems are, or even how the pilot feels. Flying is the pilot's job, and he must maintain his command and individuality. One way to do it is by telling 'em, and not asking 'em, when the conditions warrant.

Giving Our Thoughts Away

From off the Canadian coast to New York, the international airlines share a common frequency. It's Airinc, but we refer to it as "the Company frequency," since they can patch into your dispatch office and have you talk directly to the dispatcher.

Some of us call it "the Idiot frequency," because some of the stuff you hear on it is appalling. I've heard grown men, highly paid over-ocean captains, call their dispatch when the weather gets bad and say, "What shall I do?" It makes me cringe to think that a bold four-striper hasn't made up his mind what he's going to do in a weather situation.

Sometimes you call a dispatcher when things are tough and, if you respect him, ask what he thinks a terminal like New York, for example, is going to do—get better or worse? You'll ask him to get a good reading from ATC on what the delays really will be, or, if you might have to go to Boston, does Boston look like it will have delays, and how long? You could do all this ATC stuff yourself, but it would be using valuable time and cutting into busy ground-airplane traffic conversation. The dispatcher can do it by telephone and give you the results on a frequency separate from ATC's. Other times you might call him and say, "I don't like New York, I haven't enough fuel for a long hold. Washington or Montreal look like good alternates—which one do you prefer I use?" You ask this so he can decide which place is best for hotel accommodations, ground transportation, equipment juggling, and things like that. But you are flying the airplane and making the gut decisions on how it's to be done . . . to hear a pilot ask how to fly the airplane! Gad, it's awful!

Which is wandering from our subject of talking, but it is related and worth the diversion. It also relates to how you speak, which makes the man on the ground realize that he's dealing with a pro who knows what goes and will not be pushed around.

How To

How do we do that? For openers, pedantic as it may sound, a pilot needs a good knowledge of formal, correct radio usage and language. It's all spelled out in the *Airmans Information Manual*, Part I, under "Radiotelephone Phraseology and Technique." This should be learned and used! It wasn't developed as a bureaucratic caprice, but rather after years of trial and error. It's international, and has gone through the mill of the International Civil Aviation Organization (ICAO). It's good, and so is the phonetic alphabet, which everyone should know. That's the first step.

Next is personal mike technique. I've seen hundreds of pilots use mikes; thirty-six years flying two in a cockpit gives a very wide exposure to all types. What are the major errors?

1. Not getting the mike close to and in front of the mouth. Talking with the mike in some position just south of your left ear, as people do with telephones, doesn't get the message into the mike.

2. Starting to talk before pushing the button, and releasing the button before the conversation is over. This causes more repeats than you can imagine, because the complete message isn't sent and so not all received. I've heard arguments start because a pilot or controller asked for repeats and the repeater, not realizing he wasn't getting his complete message out because of bad mike-button technique, became all upset, and pretty soon two people were nasty to each other on the air.

3. Being an "uh . . ." pilot—who starts a message like this:

pushes mike button, pauses, and says, "Uh . . . Albuquerque tower, this is . . . uh . . . november one seven three seven . . . uh . . ." And it goes on like that. I flew many trips with a copilot who was one of the best I ever flew with. He came from Army Aviation, and those chaps fly well, all of them. But Al never could get over his "uh . . ." radio work. I liked him personally, and he was fun on layovers, but I disliked flying with him because his "uh . . ." technique was so damned annoying.

How to cure it? To some it's just a bad habit, but to others it's a matter of talking before thinking. One of those corny sayings you see stuck on bulletin boards applies: "Don't talk until brain is engaged."

There's a type that grabs the mike, pushes the button, and starts talking before deciding what to say. So he calls the station—that's easy: "Albuquerque tower, this is november one seven three seven . . . uh . . ." Then he's blanked out mentally, with nothing to say.

I don't particularly like the idea of writing down what I'm going to say before saying it. This business moves too fast for that, and you might as well learn at the pace required. Writing it down first isn't going to keep pace with events. It's like Ben Hogan says in his book of golf: Hit the ball hard right from the start and then get the wild shots under control as you progress. He didn't think you could start easy and build up to big hitting. Same with radio usage.

4. Talking before listening. This means listening to hear if others are talking before you start. It's more than just that; it's also listening long enough to be certain you're not busting up a two-way conversation. The ground says something to a pilot, and there's a small time lapse with the air blank as

the pilot gets set to answer. You pick up the mike in this blank space and, since there isn't any conversation, start talking and spoil the rapport between ground and pilot.

There are some terribly thoughtless and rude pilots in the sky on this score. They break up transmissions and cause endless repeats.

Sometimes, at a busy terminal and a busy hour, when you're trying to get ground control for taxi clearance, it's tough to squeeze in. It can be done, but it requires listening to the chatter closely and then, when there's an obvious break, being ready to make the call . . . and when I say "ready" I mean finger on the button, brain up to speed, and do it! It isn't easy, but it can be done, and with class.

Is It Necessary?

"Frugality" is a big, good word for radio use. Are we sure what we're about to say is necessary? Could the message be cut in half and still put the point across well . . . maybe better?

With frugality there's the examination of what you are about to say. Is the message really necessary at all? Is there another way to get the information? The simplest example we've talked about is getting weather—listen for it and don't have to request it.

The comic excess of radio use happens in glider contests. For those who haven't been exposed to this wonderful, idiotic sport, the routine is that the glider has a radio and the automobile that pulls the trailer to retrieve the glider has a radio too. In a contest the automobile, pulling the empty trailer, tries to stay near the glider. This can be over a 100-to 400-mile course—even more. If the glider has to land because there isn't lift, the ground crew can help spot a field for the pilot and give its condition: whether it's full of unseen rocks, ditches, and fun things like that; what the wind is. Also, if it's early in the day they may take the glider apart, put it in the trailer, dash back to the starting airport, take it out of the trailer, reassemble it, and start again. Since the energy crisis this operation, known as a relight, has been pretty well done away with (and a good thing, too, because some of the car driving was pretty wild).

But the "charm" of a crew following the glider is still there. Often the ground crew is the glider pilot's wife, whom he's conned into this job. The conversations between car and glider are often unique as the pilot tries to explain where he is and where the car should go. There are times, too, when lift is weak and the glider pilot may be just going up and down a small ridge, barely staying up, as he waits for thermal conditions to improve so he can climb high and be on his way.

During one of these periods I heard a pilot tell his crew, "Hold where you are."

"Roger," answered the very feminine voice. Then she added, "How long will you hold?"

"I don't know." Slightly irritated.

"Well, there's a laundromat here where I'm parked, and we haven't washed in days—do I have time to get a load done?"

"My God, no!"

Another day a female voice called her pilot:

"Ernest, I can't find the car keys."

A long pause, then a very low-key voice: "I've got them." So keys are in the air with the glider, and Ma's on the ground unable to move. This precipitated much conversation, which finally resulted in the glider pilot, who didn't want to land because he'd have to go to the end of the start line and lose soaring time, wrapping the keys in a handkerchief, flying over the field, and tossing them out. Of course the keys came out of the handkerchief, plunged to earth, and never were found. Ma got the local car dealer to make a key for her, and started out much later.

Such conversation causes angry comments from other pilots who have messages that are important. Thank goodness the glider frequency isn't one used by all aircraft for important stuff.

Using radio can be fun. The fun is in trying to see how well you can use it, how brief, efficient, and clear the message can be. Overdoing the repeat business takes a lot of time. By that I mean an original call might go: "New York from Mike Gulf Alpha November, over."

New York answers. Now they've established communication, but sometimes you'll hear the pilot, on every contact, say, "New York from Mike Gulf Alpha November, over————." "Alpha November" would suffice, because they are talking to each other and it's obvious who's who.

Sometimes—if, say, there are two similarly numbered aircraft working the same radio area—it's very important to be explicit enough so there isn't confusion. But aside from that sort of situation, the long-winded repeats, when not needed, are a waste. The mark of a real pro is the ability to sort this out and make good judgments on when he can cut corners and when he cannot. Sometimes two clicks of a mike will get it done; others it takes the entire procedure—which is what to use when in doubt, of course.

Frugality also means a lack of the "howdy" type of conversation. Talking over crop conditions, baseball scores, how cold it was back home, fishing reports, and stuff that isn't useful is a hell of an imposition on the serious folks who have things to say and ask which pertain to flying!

Don't Be Bashful

Although I've talked a lot about saving radio time and reducing messages, I don't want anyone to feel intimidated about using radio if he needs to. If there's a real problem, say whatever you want, and don't be bashful about doing it. Take the necessary time. And that's why, in normal times, we need complete radio discipline: so there is time for the one who may be in trouble, really needing the air.

Radio use—good radio use—is a matter of thoughtfulness, practice, and knowing the correct procedures . . . and not being guilty of the four bad points.

10

Emergencies

What is an emergency? Probably anything that happens which makes you scared, mouth dry, adrenalin pumping, and the big apple in your throat.

For some this happens easily; for others things have to be pretty desperate to get them in that shape.

Which tells us that some emergencies aren't emergencies at all—they are only to certain people.

There are slow emergencies and fast ones. A slow one is when you have an hour of fuel left and the nearest place that isn't zero-zero in fog is an hour and a half away. A fast emergency is when you have just taken off, are over a stand of trees at 100 feet, and the engine quits.

The slow emergency probably shouldn't have happened in the first place. What was our pilot doing in a fix like that? If he'd studied the weather and had enough fuel, he wouldn't be in such a mess.

A lot of emergencies are like that, caused by poor planning. Planning, and not being afraid to be conservative when situations are marginal, will prevent most emergencies.

But—to not leave our guy with only one hour of fuel and no place to go, just hanging there—what does he do?

He could start with prayer, because unless luck opens a field within his range, he's going to be faced with a blind landing.

He's on the radio getting all the weather possible to learn if there isn't some spot to go to. He's got his engine leaned back, and is flying at long-range cruise airspeed, which, if he knows his airplane, he didn't have to look up while up there in this pickle.

No fields are opening up. What now? Is there a good airport within range that has an ILS? If so, go over there and start making passes and learn how to land blind. Get there with enough fuel to practice approaches. Find the drift and necessary descent rates, and then fly it on the ground. It has been done. I did it once and got away with it.

Now the last ditch is that there isn't a field, it's all fog below, and the fuel runs out. Well, this puts us in the same league with that short-emergency guy with a failed engine over trees.

When an emergency like the failed engine occurs, something quick and unexpected, there's a shock moment in which the human mind, for parts of a second, just sits there and draws a blank. It is incredible that something like this could happen. It doesn't seem real.

But it is all too real. If we're emergency-planned, however, and have thought in advance about the first thing to do, we can do it almost automatically while we're getting over the shock and getting oriented to what's wrong.

First

This first action must be done! No matter what has happened you do this immediately, and above all:

FLY THE AIRPLANE!

and the first thing to do in order to fly the airplane is look at the

AIRSPEED!

Airspeed is what we live by, it's the breath of life, and if you don't have it, nothing else matters: Failed engines, fires, anything at all takes second place to airspeed.

Oh, some could argue that there might be a case, up high, when you could let airspeed go to pot while you fiddled with something else, but it's very difficult to rationalize this as being a good move.

No! The first thing is AIRSPEED.

And number one is SAFE AIRSPEED ABOVE STALL.

Your first look at the airspeed is to see if you're safely above stall airspeed and, if above, that it isn't decreasing and decaying toward stall.

The second airspeed, after we've gotten the stall speed safely in hand, is the best airspeed to fit the present situation. In a twin it's the best climb speed, and in a single it's best glide speed.

A Point Aside

But before we discuss those speeds, it is very important to go back to multiengine airplanes and talk about another speed which comes even before stall speed, if you lose an engine. It's MINIMUM CONTROL SPEED—the speed which makes certain you are going fast enough and have enough airflow over the rudder with an engine out to control the airplane. It's a speed you look for before you're in the air and an engine fails, because if you don't have sufficient speed to control the airplane with an engine out, and you're in the air, you're not a clever pilot, and in a bundle of trouble!

If properly done, the only time you'd be under minimum control speed, with an engine out, would be on the ground, and in that case you might as well cut the other engine and stop the airplane as gracefully as possible.

Now, Those Airspeeds

After that first, important stall-speed inspection and correction, we want to get at the best airspeed to fit the situation. In a twin, climbing out, for example, we want either the best *angle* of climb, if we're trying to miss an obstruction like an apartment house, or best *rate* of climb, if we are not worried about obstructions.

Any of these speeds should be firmly in your mind for

your airplane. You haven't time to look them up! So know, exactly, what speeds are recommended and best for any condition in the airplane you fly.

In a single with a failed engine, what's the best L/D speed? L/D? It's Lift over Drag—more down-to-earth, it's the glide ratio: how many feet forward for each foot down. In my glider it's about 38 to 1, for the Skylane about 10 to 1. So if I'm a mile high, I could glide 38 miles in the glider and 10 miles in the Skylane . . . no wind.

But they perform that well only at a certain speed, the best L/D speed. What is it for your airplane? It's in the manual somewhere. For my glider it's 52 knots, for the Skylane 80 mph.

While cruising, I sometimes visualize having a sudden, complete engine failure—a very unlikely happening. I've thought how if it happened I'd pull up from my cruise indicated in a careful zoom to gain extra altitude and, as the airspeed decreased, level off so the decaying airspeed would stop decaying at exactly 80 mph as I got level. It's a neat trick to do without getting too slow, and worth a practice session or two. It would go like this: Cruise at normal speed, cut the throttle, and then zoom (check traffic above) to let the airspeed drop off to best L/D speed (80 mph in the Skylane) and reach it without going under that speed—by no means go under it. See how much altitude you can gain; also check at what speed you have to begin leveling off before the best L/D speed so you hit it on the nose—that is, to get to 80 mph in a condition that is not decelerating or accelerating. How abruptly you pull up affects all this, and the pullup should not be too abrupt.

In gliders this is a standard cross-country maneuver. We

cruise, sometimes at as much as 100 knots, between thermals. Then, as we begin to feel a thermal that we want to circle in, we pull up in a zoom, letting the airspeed drop off to the best circling speed, arriving in the turn with the speed established and stabilized. It's a good precision exercise.

So first we make certain our airspeed is safely above stall (and minimum control speed for multiengine), and then get to the best speed for the condition. Those are our first FLY THE AIRPLANE chores when we're getting organized in a sudden emergency: the airspeeds.

On the Level

Next we should make sure the wings are level (unless we want to turn), and check the heading. It could better be called "direction." If it's a twin and you've just lost an engine, you don't want a wing to get way down, or the heading to start off in a curve. If it's a single you want to go straight ahead unless judgment says you can make a controlled, safe turn toward a place you want to get in.

But always airspeed is first! If you're going to crack it up, do it under control with the airspeed *you* want, and not out of spin.

So, again, first always FLY THE AIRPLANE! An unfortunate man was killed in a glider right in front of my eyes because he didn't fly the airplane. He hadn't securely latched the canopy of the glider, a big Schweizer 2-32, all metal. During takeoff the canopy came open and, hinged on one side, started to flop up and down. He was so concerned about the canopy that he quit flying the glider and reached up to try and get that canopy closed. He got into a big phugoid oscillation that finally pulled the tow plane's tail up so high that the tow plane, a hundred feet or so off the ground, was headed nose down for the dirt. Justifiably, the tow-plane pilot released the tow rope. The glider still zoomed, fell off on a wing, and spun in.

A horrible thing, especially after I talked to the Schweizer factory test pilot and learned that you could fly the glider safely with the canopy open or shut, or flopping open and shut. So if the poor guy had just flown the glider in tow to a safe height, released, and then tried to close the canopy, or left it open, he could have flown back to the airport and been alive today. Yes, FLY THE AIRPLANE first! Without a controlled airplane, the other emergencies are just conversation.

Go Fast Slowly

The second thing about emergencies is that in most cases fast action isn't necessary so long as the airspeed is proper and the airplane is being flown. That, in most cases, is the only fast action needed.

The complexities of most emergencies require a certain amount of analysis before you try to correct them. Fast action without this analysis doesn't cure problems—it only creates more.

Sorry!

One day, during my early days as a copilot, I was riding jump seat of a DC-3 while the chief pilot, Jack Zimmerman, gave an instrument check.

In that time it was thought important to feather a failed engine fast (this thinking has since changed appreciably). Jack had this pilot coming down the Newark radio range leg to cross the station at 800 feet. Just before the station Jack pulled the mixture control on the right engine to simulate a failure. The pilot immediately reached up and pushed the feather button, but he pushed it on the left engine, and that feathered! We were in a poor glider at 800 feet over the Newark meadows.

"*Keerist!*" Jack yelped as he shoved the mixture control forward on the right engine and got it going, then, after sinking lower than was comfortable, got the left one going again. Those were a bit more carefree days, so we all had a good laugh about it. But into my neophyte mind went the thought that it doesn't pay to move fast until you're certain what distance, in what direction, and what for to move.

Stories about the wrong engine being feathered are many, but the worst I ever heard happened during World War II. The early B-29s had a terrible problem of runaway props, and the training centers drilled new pilots that if an engine started to overspeed, get it feathered fast. The B-29 also had the feathering buttons clustered together on a panel. The story is of one young man taking off from a western training field, I think Albuquerque. When he advanced the throttles, one engine ran over the maximum rpm, in what probably was just a surge as power was advanced, but he jabbed the feathering button—unfortunately, the wrong one. He tried to pull it out and stop it as he realized his mistake. But, finishing its feathering cycle, it had already popped out. In his frantic tugging, he hit another one and feathered the second engine; now the airplane swerved and thundered out across the boondocks. He tried to unfeather number two and, by the same process, got a third one feathered! The airplane went across the field and over a road and broke up in three pieces before the dust settled. (They walked out!)

Actually, the entire incident was caused, probably, by too much emphasis in training on the instruction to "Feather it fast!"

Sometimes, as in this case, too-fast action will make a minor incident into a big one.

Emergencies Don't Happen Often

Fortunately, emergencies don't occur very often. I know lots of people who have never had an engine failure, and most of us have never had a fire. I've had two, one in a Connie right after takeoff from Frankfurt, Germany. The bells rang and red lights came on, so I had the copilot tell the tower we were landing back immediately. The tower started its litany about downwind, number two to land, etc.

"Tell that silly sumabitch," I said, "that we're on fire and we'll be back on the ground in sixty seconds and to get everybody else the hell out of the way!" He told 'em, we landed, and that was that.

It was a turbine failure in the exhaust-recovery system of that horrible Turbo Compound piston engine we had in late Connies; DC-7s had them too. We got it out without any damage except to a few engine knickknacks and the parts that had a hole knocked in them.

Number two was a Piper Comanche, and, while not big, it was scary. I was just west of Pittsburgh with son Rob at 8000 feet. There were snow showers below, with ceilings and visibilities from poor to lousy. Gradually, in that unbelievable way, smoke started coming back into the cabin. For a couple of minutes we swallowed hard and dry. It smelled electrical, so I knocked off the master switch, and shortly the smoke began to subside and in a little while was gone. To try to find what was wrong, we shut off all radios and turned the master back on: no smoke. So then we started a process of elimination and turned the radios on one at a time. The second one to go on, a nav-com, started the smoke again, so off it went. The smoke went with it. So that stayed off and all others came back on—no problem. The flight proceeded minus one nav-com we could easily get along without. On the ground we found that in manufacturing the radio a loose screw had been left inside and it finally bounced itself into position to cause a short, which started the smoke and smell. It was a new airplane, and could happen to any of them.

But that's a pretty good track record for almost 30,000 hours in airplanes of all sorts. Two fires, and one really only the smoke before a real fire.

Emergency Practice

The fact we don't have many emergencies is bad in a way, on a couple of counts. One is that we aren't current in handling them, and the other is that we can develop a very complacent attitude toward emergencies.

On the airline I've had hundreds of fires, but all in a simulator. It's part of an airline's recurrent training system, and it's excellent. My going through the master-switch-and-isolation routine in the Comanche, almost automatically, was really due to my airline training.

Training

We might talk about airline training a little and what their pilots think about. Primarily they train and learn about the things that can go wrong and then what to do about them. You'd think they crash one a minute, but, of course, airlines don't. Their safety record is phenomenal, and training and checking is one big reason why.

Anyone should know how to fly the airplane. This means stick and rudder, the ability to take off and land, do maneuvers, fly an ILS, and all the rest. It's like the swing of a good golfer. Being competent, proficient, and good at flying is the goal and basis for everything, and especially for taking care of an emergency or, as we might better call it, the unusual.

These basics are taken for granted, because if one doesn't do them well eyebrows are raised and job security becomes a problem. What this says is that all of us should never stop trying to improve ourselves and be better pilots in all the basics of knowing how to fly.

After that come the two aspects of handling the possibilities of the unusual: one, through knowledge, and the other, through knowing how to fly the unusual. And I'll be using the words "unusual" and "emergency" in the same concept.

As an airplane pilot sits in ground school, he learns the airplane and its systems, like the hydraulic, for example. What does he learn? What the system is, how many pumps, pressures, what it runs, etc. But he also learns, right along with the system, what happens if something malfunctions.

The first six questions on the landing gear in the 747 study guide go like this:

1. Which hydraulic systems power the landing gear?
2. What will prevent movement of the gear lever to the up position?
3. Will the gear-unsafe horn sound with the throttles retarded at airspeeds above 250 knots, when the gear is not down and locked?
4. Will it be necessary to use an alternate gear-extension procedure with loss of #4 system?
5. Is it necessary to have the gear lever down when using the alternate hydraulic-extension procedure?
6. Should the gear lever be off when using the alternate gear-extension procedure?

Of these six questions, four have to do with unusual happenings. They are designed to keep a small unusual event from becoming an emergency.

Number two question has to do with the fact that if you cannot get the gear lever up, it's because a gear hasn't tilted so it can go into the space carved out for it in the airplane, or that the steerable gear isn't centered. You can, with a release latch, override the gear, but using the override without first checking out the reasons for the problem could mean a damaged or jammed gear. If it checks out, you override and raise the gear; it was probably a minor electrical fault. If it doesn't check out, you leave the gear down, go back, and land. The point is that improper knowledge of the system, and its "ifs," could turn a minor item into a major one if someone used the override without checking it out first. Of course, this isn't anything new. Any group of pilots sitting around talking airplanes will get into these kinds of "ifs." The airplane manuals are full of them, in the sense, for example, of how you get the landing gear down by other than normal means.

So what we're saying is that the first thing to do in preventing emergencies is to know the airplane, all about it and especially the "what ifs." Besides those listed in the airplane manual, you can think up your own by constant thought and review of your airplane.

Dig for Knowledge

There are other ways to learn about that airplane. Instead of running off somewhere while the airplane's having its 100-hour check or annual, stick around and ask questions, look inside when the tin is all removed . . . being careful not to interfere with the mechanic working on your hourly time! But a little extra cost here is well-spent money.

If you get a chance to visit the airplane's factory, try to catch a coffee break or lunch with the test pilots or sales pilots and pump them about your airplane. These people know a lot more than you'll find in the books. Have a list of questions that have occurred to you from time to time in your mind or written down, so you can ask the men who know. You'll get the questions answered, as well as new ones that will develop during the bull session. It's almost impossible to know all about any airplane, but it's valuable and interesting to make an attempt, a never-ending quest.

In the Air

Now comes the part about flying and trying emergency maneuvers. It really starts with a liaison between being in the air and the things you learn on the ground. For example, you read about the emergency means of extending the landing gear. Have you ever tried it? Well, on a nice day aloft, after thinking it all through carefully and double-checking the manual, try it, see how it feels and acts. Then, if a need occurs, which might be during difficult weather, at night, and so on, putting that gear down will be a routine matter.

And this goes for all the unusuals we can do with an airplane. Do you try spot, power-off landings with your single-engine airplane periodically to learn how to make it if you have a forced landing? Do you fly a twin with an engine out, prop feathered? Well, it's all good practice.

Take It Easy

One thing the airlines have learned is that practice should be done under ideal conditions, in a relaxed, take-it-easy way. They feel the student (maybe he's a 25,000-hour veteran captain) learns better what it feels like, what the procedure is, in a slow, relaxed fashion so it can sink in.

The airlines, as I said, do their emergency practice in a slow, relaxed way, for safety reasons as well as for better learning. Too many people have been killed practicing emergencies. It's a known fact that more people have been hurt in simulated forced landings than ever have been in actual ones. People have been killed flying twins with an engine out on training or rating rides, from marginal flying. If an emergency practice session has any danger in it, don't do it!

I remember, years ago, getting an instrument check from a newly appointed and eager check pilot. It was a very windy and turbulent day. A cold front had gone through and left a winter day of gusty winds, low gray racing clouds that spit snow showers, and that down-in-the-seat, up-against-the-belt kind of turbulence. This check pilot had me with one engine out in a Connie for a three-engine landing, which was pretty interesting under the conditions. Then, on downwind, he cut a second engine for a two-engine landing! With the turbulent conditions, it was the kind of maneuver people get killed trying. I pushed the second engine back up and said I wouldn't do it.

"You've got to do it, it's part of the check."

"Well," I told him, "I've just busted the check, so let's turn on that other fan and go back home."

He didn't want to do that, so we argued awhile, and finally he "substituted" another maneuver and the check was complete.

I wasn't about to do something as silly as that. Without stopping to think for long, I can count five friends who were killed during checks or checkouts in the years I was with an airline. Which says a lot for being careful and not doing any maneuver that's dangerous. This all says a lot in favor of simulators, too.

Too Close

I was checked out on our first Connie at the factory in Burbank. It was all new then, to the check pilot and to me. Even the FAA inspector didn't know quite what to give in the check, so, believe it or not, he had me do 180° and 360° power-off spot landings with that big airplane! We got 'em done, and it was fun.

But during the check we almost splattered the Connie all over Burbank. I took off and the check pilot shut down an engine, feathered it, right after takeoff. It wasn't any real problem, although we were low and just clearing the end of the runway. But right there he shut down another engine! Things happened so fast, they were practically instantaneous. The airspeed collapsed, and in a second we were too slow and too low. There wasn't room to shove the nose down and get speed, because if we did we'd be flying through treetops and house roofs. Without even speaking, we both knew we had a serious problem.

It wasn't possible to unfeather at our low speed because, in going through the unfeathering cycle, we'd create more drag and lose more airspeed before the power came up and the prop started pulling.

Then followed a great example of two people knowing what had to be done and doing it, without words. I flew as carefully as possible out across a lot of country, slow and low, trying to build up airspeed until Jim, the check pilot, felt there was enough to risk unfeathering, which he did.

It was a very hairy experience, and when it was over Jim said, "Man, I'm never going to do that again!"

I added, "And nobody is going to do it to me again, ever!"

You Don't Have To!

It was a lesson that stuck in my mind, and I've never been bashful or embarrassed since about refusing to do a maneuver for any check pilot, company, FAA, or whoever. Living is more important than passing any check. And besides, so long as you are alive, you can argue.

Since those Dark Age days, checks have become a lot more sensible. We aren't required to make two-engine landings in four-engine jets except simulated at high altitude or in a simulator. A lot of other silly maneuvers have been cut out, too, and it hasn't detracted from safety one bit. Any check pilot who tries to make it tough by making it dangerous is an ass, and shouldn't be a check pilot!

An important point in practicing emergency maneuvers is

that if, during a simulated emergency, some irregularity really occurs, the simulated emergency should be stopped immediately and then the irregularity taken care of. It is part of our Fly the Airplane First philosophy.

I lost some dear friends just a few years back on a 707 instrument check while they were doing a three-engine-out ILS approach—really a very simple thing in a 707 or 747. But during this approach the unusual occurred, and the hydraulic-fluid level went down. Rather than recover from the three-engine condition first—that is, bring back the simulated failed engine—someone in the crew went through the emergency hydraulic procedure first, which was to shut off the hydraulic pump to save it from burning out. When that was done, the rudder hydraulic boost was lost, and the one-engine-out minimum control speed went up, because of no boost, to a value higher than the speed they were flying. The big bird rolled over and piled up—all lost. A horrible thing, and, like many horrible things in flying, a simple thing. A drilled-in emergency maneuver from school done too fast. It emphasizes the importance of stopping the simulated emergency immediately if anything really goes wrong, and moving with the brain preceding the hand.

The key in all this is that we should learn how to handle the unusual without putting ourselves in any dangerous situations while doing it. It's very important to be able to handle the unusual, the emergency, and I don't mean to discourage practice, constant practice, but don't put yourself in a position where you can get hurt doing it. We've learned, bitterly, that calm, relaxed, undangerous practice and study have actually made us safer, better emergency pilots—and kept us alive in the process!

So you don't practice spot landings with a single-engine into a little marginal field, but rather do it on a big field. Pick a spot on the runway up from the end which gives you margin if something should interfere with the practice spot landing, like the engine really quitting because it was idled back too long and carburetor heat wasn't on. Ditto with a twin on one-engine: It should be high and safe.

Practice

Now to the point of being current in unusual happenings. As I've said, unusual things don't occur enough to keep us current, so the best insurance is to keep current ourselves. It means to set aside a certain period of time once a year— twice is better, and maybe three times or four—to refresh knowledge of the unusual. The airlines do it twice a year in a simulator or airplane and then have quarterly home-study questions to boot. While general aviation isn't an airline, the airlines have done things that are useful to GA, and emergency and unusual training and recurrent training is one of the big ones to copy.

Take out that dusty airplane manual and read it over, enough so it doesn't get dusty again, no matter how well you think you know it.

The other big point in all this is that because unusual happenings don't occur very often, we tend, consciously or subconsciously, to think that they never will.

They Do Quit—Your Fault

The modern airplane and its engine is a wonderful thing. Engines, like the one in my Skylane, just don't seem to quit, and if they do, more often than not it's because they weren't treated right or, because of poor planning, they were starved for fuel.

Not being treated correctly means things such as not properly fueling and draining tanks to be certain there isn't water or dirt. As I stand around airports just seeing what goes on, I'm appalled to notice how many people drive up, tell 'em to fill the mains, come back after they've had coffee and signed the credit slip, and drive off without checking caps or sumps. I've seen line boys tolerantly half smile at the fussbudget as I get a ladder and check the caps after they've gassed it. Well, who gives a damn what they think? I don't.

It doesn't hurt to look around inside the cowling for oil leaks each time you gas it, too, even though some modern engines are cowled so tightly that all you can see through the cowl door is the oil cap and the engine name plate. But checking that cap is worth the bother too.

So if we've checked the gas and oil systems, we've made a big start toward preventing engine failures.

Counted among fuel-system problem areas is fuel management in flight, and by that I mean what the gas gauges say vs. knowing, by hours and rates, how much fuel is still in the tanks.

Most little airplane gas gauges are pathetic. They are accurate, generally, when it reads full, although we should look in the tank to be sure. But on anything except full, what they say is just rough information.

A guy ran out of gasoline one mile from the end of the runway at our local airport at Montpelier while landing on a clear, sunny day. Luckily it resulted in only a bashed-in nose wheel and some expensive wrinkles in the fuselage. But how could a guy have guts enough, or be dumb enough, to cut it that fine? It's amazing!

On big airplanes, like 747s, 707s, Citations, and others, the fuel gauges are very accurate. They are capacitance types that measure the electrical resistance of the liquid in the tank and from that tell what's in it. It's really a comfort to look at the gauge and know what's in the tank, and I wish they could make them inexpensive enough for all airplanes.

But even with these gauges, the airline crews keep a fuel-use record: a record of how much they've burned by the rate per hour vs. the time, and that deducted from the amount at the flight start, on a running, current basis.

This is easy to do, and it doesn't take a mathematical genius. Mostly a note on a piece of paper is enough. But time and rate are the important things to tell what's left in the tank, and not the indication of a bouncy fuel gauge. Even the good fuel gauges fail now and then, so then where are you without a record?

Dipstick

I also wonder what ever happened to the old idea of sticking a tank. Airlines still do, even with their fancy fuel-reading systems. It's not a bad idea to take the time to make a fuel stick. Stand the airplane level, empty the tanks, and then refill them a small amount at a time—say, enough for a half-hour's flying. After adding each amount of fuel, put the stick in and see where the level is for that amount of fuel and then make a mark on the stick. By progressively doing this, we can finally have a fuel stick. When we don't remember just what was left in the tank, we can stick it and know for sure. With fuel shortage problems, there may be lots of times when fuel isn't available and we want to know if there's really enough in there to fly to the next field that has some.

A stick is a good thing. So using it and then keeping track of the amount burned and remaining will go a long way toward not having an engine stop because of "fuel starvation," as the accident reports put it. And with it, check after fueling and each time before the day begins for dirt and water.

That comes close to taking care of fuel problems so far as the airplane's concerned, but it doesn't take care of all the fuel problems created by the guy flying the airplane. Most failures involving fuel are not running out of fuel, but fuel mismanagement. Taking off on an empty tank; ditto for landing. And there have been many cases where a pilot landed "out of gas" when he actually had lots of fuel in another tank! Airplane fuel systems are often designed in tricky ways that invite pilots to make mistakes, so it behooves us to know, in great, intimate detail, the fuel system of each airplane we fly and then set up operational procedures that will prevent the tricky system from doing us a bad turn. And whenever an engine quits, a check on the fuel supply and where it's coming from is an early, quick action.

How Much Fuel?

How about the pilot who's willing to fly until his fuel is almost gone? How much should he have for reserves? Well, personally, I've always been a sissy about reserves. I like a lot. I've not always had a lot, but if I didn't it wasn't because I hadn't planned for more. In earlier days of ocean flying there were many hours of sweating it out as the groundspeed and fuel both got low. I've squirmed many layers off the top of seat cushions in such situations. A couple of times I've turned around and gone back. I can recall a DC-4 crossing from Shannon to Gander. We got slower and slower, and the weather at Gander got lower and lower. Finally it was evident that we'd arrive there with bare minimum fuel and no alternate capability. I felt foolish doing a one-eighty over the middle of the North Atlantic Ocean, but I did, and went back to Shannon.

(Which, while it doesn't have anything to do with this, reminds me of a night in a DC-4 when we were going very slowly into a monstrous head wind. The navigator suddenly appeared in the cockpit and, looking rather wild-eyed, yelled to me: "My God, we're standing still!" I knew things were tough, but couldn't visualize 'em that tough. At the navigation table I went over his figures with him. We discovered he was using the wrong date in the Almanac and his sparse star shots were all in error. We were really doing 90 knots.)

I turned around in a B-17 exactly halfway between Honolulu and San Francisco. That one was simple: Everything at midpoint said we'd be dry-tanks before the California coast.

Flying my Cessna Skylane, I'm almost a half-tank man—meaning that when it gets down to half-tanks, I begin to get organized for landing. I've got the big tanks, with over six hours' range.

Now, I don't always land at half-tanks, but I can assure you that legal FAA reserves are minimums that I don't want to get down to. Remember, an FAA regulation is designed for the minimum, the least. When I have bare FAA minimums, I have a worried look on my face unless things are in excellent shape.

Any minimum is directly related to weather and available airports. If the weather is sour, it's getting dark, and so on, we're going to want big reserves that'll get us to a safe alternate. If we're headed into noon and a beautiful day, it's a matter of what you like to have toward the bottom of the tank.

But what is that safe bare minimum for daily operation? To me it's anything over an hour that I can feel confident of having. Confident means an amount I know precisely, derived by keeping track of the amount burned, deducted from an amount I knew I had learned from full fuel, or dipstick measurements, plus previous recordkeeping. Eyebrows may go up at the figure of one hour plus as being too much, but life is much more pleasant when it's relaxed. And running out of fuel is one of the more stupid things to do wrong in an airplane. I still can't get over that guy who ran out, on a sunny day, one mile from the end of the runway!

There are temptations to stretch fuel. You're almost home, it's late in the day and you don't want to stop and perhaps get into night flying. But these are facts of life, like taxes, and sometimes one has to pay the price to be safe . . . and smart.

It Isn't Always Easy

The only time I ever overflew Paris in a 747 was on a foggy morning, and it was one of the toughest decisions I've ever made, even though a very undramatic one.

I arrived over Paris with 300 passengers at about 8 A.M. It was socked in solid. My alternate, Marseilles. I had pretty good fuel aboard, so I held for it to improve. I went round and round up in the sunshine, on top of a solid white fog bank over Toussus-le-Noble, which is near the Palace of Versailles.

After an hour Orly hadn't budged. The visibility wasn't even 100 meters, and our minimum for even a look-see is 200 meters, with 400 meters required for landing.

The situation dragged on. I knew it would open up, everything said it would: low fog cover, clear above, the day young. But as I approached my minimum fuel, I was faced finally with the fact that if I didn't get going I'd be burning into my alternate fuel and not have Marseilles capability. I knew Orly would open up before my fuel ran out—but what if it didn't? There was a 1000-to-1 shot against it, but can you imagine anything worse than a dry-tank 747? Even with my glider rating!

So I peeled off and went to Marseilles. Orly opened up before I landed there. I got fuel—a two-hour chore—and went back to Orly.

I still grind my teeth over that one. Should I have been

gutsier and stayed around and landed, or did I do the right thing? Of course it was right—it was safe, and that's what counts—but I screwed up a lot of people and made them late with missed connections and all the rest. But as time mellows the decision, I guess it doesn't matter any more to the few delayed people.

So going to Marseilles and all the other things we talk about here are part of planning to avoid emergencies. If we plan enough to avoid emergencies, we'll probably have few, if any—ever.

They Do Quit—Their Fault

Occasionally engines do quit from normal quittin' reasons —failed parts, etc. I saw a friend of mine have a cam ring break when he was on top of a loop, not far off the ground, right over the airport. Breaking a cam ring stops all the valve action, and that stops the engine—immediately! He did the sweetest roll-out and landing, without a ruffle, that you ever saw. Lucky and good. Warren Noble was his name.

Engines that can quit brings us to the question of what kind of country one is willing to fly over on one engine, what kind of weather, and in the dark or not.

Tough questions. Being old-fashioned enough to have flown airplanes when engines quit a lot more often than they do now, I am probably more conservative than the modern pilot raised on flat opposed engines. The new ones rarely quit, but all that does is raise the odds and make it a better gamble.

So what kind of a gambler are you? I'm a poor one, because I lost too often when engines quit regularly. Now, the modern fellow who has never lost, and has these new good odds to play with, probably will gamble much more than I do, and I can see why. But sometimes even the long odds go against us, and we lose the whole bundle.

So, considering that one can lose on a long shot, I'm a little shaken the way I see people go out across impossible-landing country on one engine: trees, rocky mountains, and the kind of terrain where landing is a certain accident.

I fly in mountain areas a lot on one engine, but I don't mind a detour that will keep me over a few places to get down, even if it's quite a few extra miles. And that goes for trees, water, and the rest.

Actually, I'm more comfortable over water, because it isn't that bad to land on, unless the seas are running high. Over water I carry life-saving equipment, flotation vest, and a raft if it's far. And I make very sure that I'm on a flight plan so somebody knows I'm out there.

Let Them Know You're Out There

Being on a flight plan is good most anywhere we fly, and it's especially true over poor terrain. It doesn't have to be an instrument flight plan—lots of times that just means extra miles. VFR is good enough if conditions warrant it. But having someone know you're out there, someone concerned about you if you don't show up, is very important.

This gets even more important in winter, when it's cold. Where I live in New England, I file everywhere in the winter, no matter how good the weather. If you land out somewhere and it's below zero, you don't want to have to wait very long for somebody to come by and pick you up!

Emergency Gear on Board

Along with all this I carry emergency equipment in my airplane. In the Skylane it's a small red nylon knapsack. In it is water—about 5 quarts except when I'm going to fly desert country, then it's a lot more; also signal flares, signal mirror, a first-aid kit (not a store one, but one put together myself after talking it over with a doctor friend); food concentrates that have indefinite shelf life, like pemmican, bacon bars, tropical chocolate—enough to last me a few days. I have a space blanket, matches, snare, fishing line and some hooks, one of those wirelike saws that wrap around their own handles, and a few other knickknacks. It sounds like a lot, but it's only a package about 15″ by 14″ by 9″ and weighs 13 pounds. It sits back in the corner of the baggage compartment and is never removed except for periodic checking and bringing up to date. The water is in tough plastic containers that never leak. In the winter the water in them freezes when it's in a subfreezing hangar or tie-down, but it unfreezes and doesn't bother the containers. More than once I've felt a lot more comfortable knowing that stuff was back there.

Best Friend

In addition, I always carry a pocket knife. Not the drawing-room version, but one of those red Swiss kind with a few gadgets—not the fifteen-gadget job, but one that fits in a pocket without being heavy.

TWA's emergency department and its training was started by a wonderful guy named Bill Davis, or Sailor, as we called him. His lectures on emergency evacuations and survival were dramatic, gory, and fascinating as well as instructive. He knew his stuff and tried to make certain you knew yours. One of his pet items was the pocket knife. He'd look you in the eye and say, "Do you carry a knife? A good one like this?" And he'd pull out his and show its healthy, sharp blade and say, "That little bastard can save your life, carry it!" And you did, and I always have.

In the Dark

Now back to single-engine flying. Night flying has a certain hazard: If the engine quits, chances are you're in deep trouble unless there's an airport underneath. I fly at night a little bit now and then to get home late in the day, but I'm never comfortable and try to avoid it. Lots of people just go off anywhere they like and it doesn't seem to bother them. My son, who's been raised on modern engines, does a lot of night flying in our single-engine airplane that his old man won't.

And on Instruments

Now about instruments on one engine. Well, you have to fly if you're going to make the airplane useful. I can rationalize a lot of single-engine weather flying because most of the time there's some sort of ceiling under you, and in a slow-speed descent at a safe but near-stall mush, you'll probably see something as you come out of the clouds—at least enough to land in or stack it gracefully.

I don't fly single-engine instruments when I have to cross big areas of zero-zero or very low ceilings under me. And I don't feel comfortable over mountainous country when only the bottom of the valleys have any ceilings. But eliminating these conditions—and by "eliminating" I mean staying on the ground if it's that way—one can still do a lot of weather and instrument flying.

The big point is that when VFR is marginal, especially in mountainous terrain, you're much safer being well up above the ground on legitimate instruments. The easiest way to break your neck is by trying to stay VFR, poking around in reduced visibility with precipitation smearing the windshield. This is not just an opinion; the record proves it. Plowing into something while flying VFR in poor weather is one of the most popular accident methods going.

One trouble with IFR is that ATC and the airways' structure don't give a damn where they send you in relation to terrain. Try coming out of Kennedy at New York. I've done it in the Skylane a number of times, and I'm over either the ocean or downtown Brooklyn. Believe me, I'd rather land in that sewage-ridden ocean than among the buildings of Brooklyn.

How to improve it? Well, one way is to file a route you think will take you over the most hospitable country. If ATC will not give you that route, ask again, explaining that you're one-engine and don't want to go over hazardous terrain or water.

Another thing we might do is to get ATC to realize there's a problem and try to take care of it in airways routing. This means letters to Washington and the FAA, plus asking your pet organization, like AOPA, NPA, etc., to fight for it in Washington. If we could just put a note on the flight plan that indicates we're single-engine so ATC would automatically give us a "softer" single-engine route, it would be pretty nice . . . and why not?

What to Say and When

Close to ATC is the question of communications during emergencies. Despite the flying movies Hollywood puts out, it isn't always necessary to grab the mike and yell to the ground for help. As a matter of fact, most times the ground adds to the workload.

I've had more than one case, and heard lots more. A guy has an engine failure and tells the ground. Immediately a flood of questions comes back: "What is your position? Which engine is it? What kind of failure?" And on and on, depending on the guy on the ground. These questions seem pretty inane to a pilot who's busy cleaning up his emergency condition and trying to decide if she'll fly okay and if so what's the best next move. Believe me, it isn't the guy on the ground who has the answers. So if you're kind enough to announce your problem, or think you must because you might interfere with ATC, etc., and then a flood of questions comes pouring at you, don't be bashful about saying loud and clear, "I'm busy, don't bother me!" If you're too busy to say even that, just ignore the ground.

The modern system seems to breed the idea that we must keep in contact with the ground at all costs. Well, during the first seven years of my flying life I never flew an airplane that had any kind of radio in it. I went coast to coast quite a few times, flew to Cuba, Canada, and Mexico, and, by golly, made it without one squeak of radio!

Of course, today we have a more complex system and there are times we have to use radio, but we use it a lot more than is necessary.

Funny as it seems, our brother pilots are often serious offenders. More than once I've heard an airline pilot say he'd lost an engine out over the ocean and tell Oceanic Control what he wanted to do. Immediately half a dozen wits in other airplanes start calling him, asking if they can help and what is his problem. After the first guy asked him if he needed help, and he said "No," there wasn't any reason for anyone else to ask again.

Ask If Need Be

We shouldn't have any compunction about asking if we need something, and that goes from big problems like being lost down to little ones like the fact that you just don't understand the clearance that was shot at you fast by an ATC controller imitating a tobacco auctioneer.

But also we shouldn't have any compunction about not using the radio until it's convenient. As I said in the beginning of all this, FLY THE AIRPLANE. You cannot fly a mike, and it will not keep you from spinning in, put a fire out, get you over the trees with an engine out, or anything else.

So when we speak of emergencies we think of communications as something for our help and convenience and, if necessary, to warn others if we're a menace to navigation. But the last thing to worry about is grabbing the mike; grab the airplane first!

Other Things That Quit

To keep emergencies under control, we must also consider the possibility of electrical and radio failure. We may not have to communicate, but we sure have to navigate if we're on instruments. On the many airplanes with only one alternator, this becomes a real possibility. Airplanes having dual alternators aren't likely to have total failure, but it's possible. Electrical systems are complex and strange things, and events occur to knock them out that engineers never thought of. The 707 electrical system was designed so that it would be impossible to lose the entire system—no way. But someone finally did it. It was mismanagement plus a failure that hadn't been covered in training books. It's in the books now, and again it's impossible to lose it all in a 707 . . . until the next time.

Which, simple as it sounds, means: Always have a working flashlight with fresh batteries in the airplane. (Two wouldn't hurt a bit.) It can be the most important gadget you ever purchase.

With all the wonderful solid-state battery radios around,

it is well worth while to have one on which you can at least receive VHF when the ground sends you information after you've done that little "I've got a problem" triangle the *Airmans Information Manual* has listed. How extensive you want this radio to be is your choice, based on weight, money, and so on, but just to be able to receive VHF is a big step toward making a big emergency a little one.

Redundancy

I've always been a firm believer that one should have a gyro instrument that's independent of the normal power source for the instruments. If, for example, all the instruments are electrical, I like to have at least a Turn and Bank that's vacuum-driven. A battery-powered gyro could do it too. But, whatever, there should be some way to prevent one failure from taking away all ability to fly the airplane on instruments. Can you imagine being on instruments with no instruments! That's why I like an alternate source. I've got both electrical and vacuum in my airplane.

That's the way we think about emergencies: proper planning to prevent them, knowledge to help keep an unusual happening from becoming a big emergency, and keeping our chance-taking level as low as possible. Like the fellow says, "It takes thousands of nuts to hold an airplane together, but only one to take it apart!"

11

Sensory Illusions

Part of a pilot's character is his store of knowledge. I don't mean what the Federal Air Regulations say, or what's required to pass an FAA examination, but rather more basic stuff that becomes instinct. This knowledge is the kind that makes him adaptable to different environments and situations. It's his automatic flying know-how. We're going to talk about various things that make up this know-how, and we'll start with sensory illusions.

Do You Really See It?

Your eyes don't always tell the truth, and what you see isn't always accurate. This came to my attention in 1951, when I first met an interesting pilot by the name of Prosper Cocquyt, at that time chief pilot of Sabena, the Belgian airline.

Cocquyt was an old-timer, and had been chief pilot since 1927. I met him in Bermuda at a Flight Safety Foundation symposium and he told me a story of flying a Fokker trimotor between London and Brussels in 1931.

He was flying contact at night. It was the time when instrument flying was still a vague idea, and not used. Cocquyt, a strong man physically and mentally, spoke English with a thick accent.

He told of passing Dungeness, a point on the English coast across from Cape Gris-Nez. "I made a turn around the lighthouse to fly towards Folkestone. I was flying below the clouds about 150 meters [450 feet] . . . light rain, visibility one or two kilometers [about a mile] . . . After the turn I met certain difficulties in following the coast, as my aircraft developed a tendency to turn to the right, but I didn't attach any importance to this at the time. I descended somewhat to improve my observation, when the copilot suddenly pulled the wheel, shouting that I was very low. I told him he was mad. But he saw how low we were by the reflection of the green wing-tip light on the water! He was right! He saved my life. I gave that fellow a cigar!"

What had happened was that Cocquyt was using the lighthouse off to his left for reference. His right wing was down, because there wasn't any visibility to show him where the true horizon was. The bank gave a tilted false horizon, and, looking out the left side, it seemed as though he was higher than he actually was.

The same sort of illusion can also occur with a light ahead

as the only reference. Lift the nose and it raises the horizon in reference to that lone light. This gives a false sense of being higher than you really are. It's a great way to land short.

Cocquyt checked on other accidents and found a similarity in that two night accidents in poor weather had been by hitting on the right wing. He became very interested in sensory illusions and made a deep study of them. His papers won acclaim and many awards.

There has been much work done in this area since, with interesting things to talk about. Let's start with being on instruments, no outside visual reference. We all know that without outside reference we cannot determine our position in space and so can't fly except by instruments.

Now, can we be a little bit visual, not quite all on instruments? Yes, of course, but the big, big point is that when we see only a little bit, we see only part of the picture, and even though we see mother earth, day or night, a little bit is not enough reference to fly the airplane or, to be fancier about it, maintain orientation in space.

Want to prove it? Take off at night from a dark airport that doesn't have any lights off the end of the runway, as we did from Shannon southwest toward the river. Everything is fine until you pass the airport boundary and go out over that dark river; then it's just as though someone had pulled a sack over your head. There isn't any reference. On this takeoff we'd want to be ready to go on instruments as soon as the runway's end went by, even though it might be clear with stars above.

If you ever make that experiment and try to fly in a black

void, be sure to have someone along who's an experienced instrument pilot and can keep things under control.

Now, just one light or a close cluster of lights isn't any better. The lights are a point out in space, in that big black sphere, and will not tell you if the nose is up or down or the wings level; the light tells only direction in its most simple form—whether you are flying toward it or away from it.

If we're low and trying to fly by reference to those lights, as Cocquyt did at Folkestone, we can get a wing down easily, have a false sense of attitude and altitude, and get into a lot of trouble.

Daytime can be as bad. Fly in snow, ground all white, visibility poor; a few land objects, leafless trees, a building go by, but they aren't a solid reference. The white on white makes it possible to have that tree, building, or whatever not useful as a horizon. A wing down isn't detected by reference to those objects in that white world—or the nose, either. This can happen in thick haze or smog lighted up as one mass by the sun, too. There are lots of conditions, and they are serious.

About that business of having the nose up without knowing it, and so getting a false sense of height: With the nose 1° higher than normal, a light 3280 feet ahead will make your altitude look 57 feet higher than it is; if the light is 6500 feet ahead, your altitude will be false by 115 feet. That's for 1°!

While Cocquyt brought sensory illusions strongly to my consciousness, I'd experienced some, and I guess some sort of guardian angel kept me out of trouble before I knew what illusions were about.

A Black Night

At the end of World War II, I was landing a B-17 on Johnson Island in the Pacific. It was night, and just before I got there all the lights went out. I didn't have fuel for any other place, so I had to land.

They put a jeep with some portable lights on the end of the runway, and I made an approach. It was about the toughest one I ever made. Every time I tried to get lined up, the lights on the ground would go all over the sky. If I put a wing down, they zoomed into the sky on the wing-down side, or dropped away on the other. A small change in pitch and the lights would go high or low. It was a mess. I finally made sort of an instrument approach. I'd look at the lights and decide which way I wanted to turn to get lined up. Then I'd go inside the airplane and fly instruments, heading and altitude. After a few moments I'd look out to see how the lights were coming. Then back inside again to make a turn by instruments, then outside again to see if it was enough or too much, then inside again. By this in-and-out method I finally got the runway ahead of me. After that it was a matter of descending at a rate, on instruments, with a close check on altitude and looks outside until I finally got her on the ground.

Good Weather Too

There are other illusions besides not being able to orient yourself in the sphere you fly in. There are clear daylight conditions that can baffle you—like landing straight in to a runway that slopes up. You'll undershoot. Straight in toward a runway that slopes down, you'll overshoot.

In the first case, a runway that slopes up and causes an undershoot, the glide path seems steeper than it is, which makes you try to fly a lower approach. The opposite happens with a downsloping runway—an approach seems lower than it is.

Approach lights make you seem higher than the real altitude, and a runway appear closer than it is.

You'll think you're higher than real altitude when flying in darkness, haze, and smoke. A lot of this is because there aren't shadows. Shadows aid depth perception, and without them it's difficult to judge height.

Sometimes runways aren't in contrast to the surrounding terrain; their surface looks the same as everything around the airport, and that makes it difficult to judge height.

Flying in rain gives illusions. There's an error in refraction that makes the horizon appear to be below the real horizon.

You can combine illusions. I mean the visual ones we're talking about and mental ones. Mentally we can con ourselves into thinking we are at a certain altitude just because we want to be at that altitude. We have subconsciously convinced ourselves that we are at one altitude when we're at another. So we cannot let fixed ideas stay fixed without constant checking to see that we are where we think we are.

Like Cocquyt's light at Folkestone, an entire city's lights, all bunched together and perhaps on a slope, with the airport near the city, can give illusions of altitude and cause a poor approach.

Only Part of the Story

It's evident that there are all kinds of sensory illusions in flight, and probably some we haven't discovered. But the accidents that seem to be caused by them are too frequent. The number of worldwide jet-transport undershoot accidents is close to fifty in the last eight years! Pretty impressive.

There are lots of accidents on approaches during bad weather that we shrug off as somebody trying to get in when the ceiling was too low.

Well, that's a little too pat, and especially so when you discover that a lot of "bad" weather accidents happened when the ceilings weren't real low, but actually in a range between 400 and 1500 feet. Think about that for a while and the picture becomes quite clear.

Our pilot shoots a low approach, then breaks out at 800 feet. The visibility can be bad or good. Good visibility may be worse. But he breaks out, sees a light or two, and starts flying visually. The bad part is that he is in a condition which gives him only part of the story. He is in half instrument flight. We don't call it that, but maybe we ought to. And half instrument flight is dangerous flight. We must decide which method is keeping the airplane right side up, at the correct altitude, and headed in the right direction. Are we doing it by instruments, or are we doing it visually? If we're doing it visually, we have to have *complete* reference that will keep us oriented and at safe altitude. If we don't have complete reference, then we're still on instruments and should fly by instruments even if we can see some things outside.

I can visualize our pilot breaking out, seeing a few lights, and then, because of a sensory illusion, running into the ground or a hill. This can be aggravated by wispy scud floating over that ground or hill.

But what about the sensory illusions? We cannot know every one, then look around and say, "Ah, today we could be had by sensory illusion number two, so we'd better do this to prevent it." No, that doesn't work, but we can develop simple knowledge and methods in our flying to prevent all illusions from being disastrous. After all, fifty undershoots in eight years in over millions of jet transport landings means a lot of other guys were doing something right and keeping out of trouble. What's the secret?

Being aware of the problem and realizing that we have instruments to keep us rightside up and tell how high we are. The guy who breaks out on an approach and sees something, and is smart, doesn't deviate from the approach procedure. If there's a glide slope, he refers to it until within the field. If there's a VASI, he flies that in the same way.

Without glide-path controls, as on an omni approach, we hold the minimum altitude until the runway is close and we know that illusions, scud, or whatever cannot spoil our vision. When we decide to leave our last altitude and go for the runway, we should use the oldest trick in the business to see if we're getting too low or too high: Simply, as Wolfgang Langewiesche told us twenty-five years ago in *Stick and Rudder*, watch the spot on the runway you're aimed for. If it remains steady during your approach, you'll land on it. If you go below, it will appear to rise and you're undershooting. If it goes down and you appear to climb above it, you'll overshoot. This is a simple thing, but one we can do very well. I've used it on everything from gliders to 747s, and it works. You can practice on every landing and, as a by-product, make more precision landings to the desired spot every time.

So we come in on a non-precision approach, break out of the clouds, *and then hold* our minimum instrument altitude until we see that it's a safe point to start descent from. Then we do it aiming at the spot on the airport where we want to land, for reference. If that spot doesn't climb, we're not going to get too low.

A Potent Point

There's an obvious point that's so simple it's almost insulting to mention it, but we will anyway. It's like this: We're on an approach and see the runway or airport lights, but as we approach the lights disappear. It's obvious that a hill or some other kind of terrain is between us and the airport, and the immediate action is to pull up. It could be scud, but that calls for getting altitude too.

Long Approaches

Sensory illusion can get you on a sparkly clear night on a long straight-in approach. You just don't know what's out there and how those lights way ahead are making you fly where you don't think you are. The answer? Don't make straight-in approaches, especially in hilly terrain. Instead, keep a safe altitude and make a pattern around the field. There's nothing wrong with old-fashioned circling the field, going downwind and base. It's important not to make the pattern big. Keep it in tight so you will not get wide and maybe out over hills, towers, or whatnot out where it's black and illusions work.

Instruments—The Only Truth

The big point is to keep reference to instruments, even VFR. Even when you have stopped flying instruments and are visual, a glance at the altimeter and horizon is security. We have instruments, and they are an integral part of flying. A good pilot keeps them within his scan under all conditions. If Cocquyt had had instruments and the know-how to use them in the Fokker, he'd never have gotten too close to the water.

Fly a glide slope if there is one on any and every approach. If you're on it or above it, there isn't any trouble. Kill the instinct to go visual too soon. Don't make straight-in approaches over long stretches, and know altitudes that are safe. When in doubt, that altimeter and horizon can save the day.

12

Pilot in Command

As on a ship at sea, there must be only one person responsible for command: the captain. When you are flying an airplane, you are captain, be it a Cessna-150 or Boeing 747. The FAA spells it out in Part 1 of the Federal Air Regulations, the section called "Definitions and Abbreviations":

" 'Pilot in Command' means the pilot responsible for the operation and safety of an aircraft during flight time."

Then again in part 91, "General Operating and Flight Rules":

91.4 Responsibility and Authority of the Pilot in Command.

(a) The pilot in command of an aircraft is directly

responsible for, and is the final authority as to, the operation of that aircraft.

Those are beautiful words that we should cherish, wrap up tight in our thinking, savor, and then realize the responsibility they place upon us.

In this day and age, with everyone trying to move into the cockpit, the special phrase in that group of words is "and is the final authority." Keep that bright in your mind, and whenever someone tries to move in, remember who the final authority is . . . you!

This section of the regulations requires constant vigilance by all pilots, solo student to million-miler. The vigilance is needed to never allow this authority to be removed from the regulations. If they ever attempt to take it out or dilute it, we should grasp our battle standards and march on Washington!

It has to be one person, because a flight is the bringing together of many pieces of information and skills. To make a flight successful these pieces have to be related to the entity, because they affect each other. The only place this puzzle can be put together is in the left front seat, the seat of command. And the only person who can put it together is the commander, you, the pilot.

This command falls down when the pilot lets others fly his airplane. He lets others fly his airplane for two reasons:

1. He's naturally timid and allows people to push him around.

2. He doesn't know enough to have confidence in his decisions, so lets others make them for him—a disaster!

Let's take the first one, timidity. It doesn't solve the problem by simply saying, "Don't be timid." But the way not to be timid is to know the subject so you aren't overcome by indecision, which makes you timid.

How does that go with a pilot of fifty hours in a Beech Sport on cross-country? The answer is that he flies within himself. How?

First, it's undoubtedly a VFR flight, and planned that way. So his biggest weather worry is "Will it be VFR and stay that way?" And this is where the first timid, milquetoast mistake often is made, as he says to the FSS or weather-service briefer, "Can I make it VFR?" or something like that.

Now, that briefer cannot guarantee VFR. The pilot can hear from the briefer that it's forecast VFR—or, better, the pilot can read or listen to the forecasts and see for himself if it's forecast VFR. Actually, that's all the briefer does, and essentially what the weather man does in a National Weather Service (NWS) station. The pilot may also look at the actual weather reports and judge if it's VFR, just as the briefer does.

In a pilot's training he learned the mechanics of reading forecasts and sequences, so he can do it, or should be able to if he's learned his lessons well.

If he reads all this himself or listens to it on weather broadcasts, he is getting a better picture and feel for the weather than the average broad-brush picture a briefing gives. And, most important, digging out this information on his own may alert the pilot to the possibility of changes so that if the weather goes sour it will not come as a shock. With this possibility in the back of his mind, he'll have some alternate action planned, no matter how remote. By digging out the weather himself, he'll have a more intimate idea of the day and its conditions; he'll be making the judgments, not some groundling.

In this regard I want to say something that is tremendously important for any pilot. It is this: A pilot must be a meteorologist—not able to draw maps, perhaps, but a constant, never-graduating student of weather. Weather is not exact enough, nor will it be in our lifetimes, for a weather office to tell pilots exactly and accurately what it is and what's going to happen. Much of the time it will be different from forecast and the briefing. A pilot must understand the strange, capricious, and often dramatic way weather does things other than those expected. Unless a pilot is a student of weather, he cannot cope with these changes, which really are part of the normal in flying. After forty years of flying, there are still weather books scattered around the house where I can browse in them when I have a few minutes. I have a barometer, wind vane–anemometer, and thermometer. I listen to forecasts and then observe what actually happens and try to relate the whys and hows. Besides being useful, it's fascinating, and I'm constantly learning something new.

The Pilot on His Own

It's important for the pilot to get his own picture. A flight cannot be flown in the FSS or NWS office. There one gets an indication of what the weather probably will be, but what it turns out to be, and does to one's flight, becomes real only after the airplane is in the air. And since it's often different from the picture one got before flight, a pilot must be prepared and flexible enough to cope with the differences. Advance weather information is one of the considerations in planning a flight, but it isn't fact until after the flight is over. In his kit of parts for the flight, the pilot must have a plan and conviction to watch and obtain weather as he flies so he can change his flight as the weather changes. What the pilot sees out the window is the moment of truth, the one that counts, and the pilot in the air looking out of that window is the only one able to see the real thing.

For instance, take the simple case of a pilot starting on a VFR flight with VFR forecast. En route he finds that it isn't VFR over higher terrain. What to do? If he's our fifty-hour pilot, or not instrument-rated, he turns around or picks an airport and lands. If he's an experienced instrument-rated pilot, he gets on the radio and files a flight plan—provided, of course, that he's kept up with the weather and knows what's going on ahead and at the destination.

The big, important point is that in both cases the pilot was in command and made command judgments within his abilities.

In either case he wasn't timid about a decision, because he had sufficient knowledge to have confidence in what he was doing, could do, and couldn't do.

In the VFR case he knew it wasn't possible for him to fly instruments, so he could not continue. It was an annoying decision, but certainly not a dramatic one—so long as he didn't keep sneaking along when his inner sense said he shouldn't because he really couldn't see enough or wasn't high enough to miss all terrain, TV towers, and junk sticking up in the air.

Other Factors

The "pilot in command" is also concerned with the mechanical condition of the aircraft. A minor item isn't working correctly: Should one go or not? The pilot may not have the answer, but he should be able to talk with a mechanic and get it. The important point is not to say to the mechanic, "Is it okay to go with it that way?" Rather, ask what the implications of the malfunction are, how it (or they) might cause further failure. Then the pilot decides how it would be without the item, what further failure would mean to the kind of flight he's going to make, how he feels about it. If it's VFR, he might go without a horizon, for a simple example, which he'd never think of doing IFR. These factors only the pilot can put together for the final judgment.

A command pilot isn't pressured by people into taking an overload because he wants to be a good guy, or because someone says, "Ah, it'll get off okay, I do it all the time." He makes decisions based on what he knows, and if he doesn't know enough, he finds out.

One Is Best

A great danger in decision is two pilots flying together. On the airline we say that two captains flying a flight, instead of the usual captain and copilot, make only half a crew.

If indecision ever is going to sneak into a cockpit, it's

when one pilot wonders what the other thinks or, being a timid soul, doesn't want to make a decision until he's asked the other pilot if it's okay.

A serious problem of two pilots is that the pilot flying may make a decision and then the other pilot says something like "Don't you think it would be better if————?"

We run into that on the airline; the most troublesome time occurred when the 747 first came into being. Because it paid a copilot more than what he'd get flying as junior captain on a 727 milk run, many junior captains bid to fly the 747 as copilots.

Most of them were fine, and recognized that they had decided to fly copilot and would be in fact copilot. But a few resented the fact they were flying copilot, thought they were better than the captain, and constantly tried to "fly" the flight, which is very bothersome to a captain.

A copilot is on the airplane for four reasons: to help the captain with the workload, take over if the captain becomes incapacitated, to bring to the captain's attention anything that's unsafe, and—what some forget—to learn.

Some things copilots do that are annoying include talking on the radio too much, starting arguments with ATC, moving gadgets in the cockpit without keeping the captain posted or checking with him first. One of the most disconcerting actions is to have the radios set up for navigation— say, one omni ahead and one on a station to the side for a cross-check—and, seeing things not working out, realize that the copilot has reset an omni without asking about it first. Arms have been broken over that one!

I flew with a copilot in early 747 days who was like that.

He was an excellent pilot, and knew it to the extent that he believed he was the best in the sky. He made flying very unpleasant for me until we had a chat which went something like this:

"Look, you're here to help me. If you see anything that's unsafe, I want you to tell me immediately. Aside from that, don't heckle me. I'm captain of this airplane, and I'll make the decisions."

Tougher words than I like to use, but they worked, and after a few hours of sulking he got back on the runway and things went much better.

The point is that when two pilots, perhaps very equal in experience, fly together it should be understood that one of them, picked in advance, is the commander of the flight, the other copilot. If the copilot sees something dangerous—like getting too low, missing an ATC-assigned altitude, and things of that nature—he sounds off. Beyond safety matters, he keeps quiet.

There isn't anything wrong with the captain's discussing an action like going to an alternate, a mechanical problem, the weather, anything, with the copilot, and I've done it many times, *but* the discussion is only for information with which the pilot in command will make his decision! He shouldn't be backward about making a decision that will be different from the one suggested by the other pilot. The copilot may be in the airplane with the commander, he may see most all the things affecting the flight, but he doesn't know how the pilot in command feels and how he has related all the factors, some of which the copilot may not fully understand.

Alone

A good captain never abuses his privilege of being captain by making irrational demands, acting rude, or insulting the dignity of his co-worker. The fact he commands brings serious responsibility in this regard, but it need not dilute the command.

When a pilot opens the throttle to go aloft on his first solo, he becomes a "pilot in command." With command he is thoughtful and balances all factors, but when he makes his moves they are done with firmness and conviction, and he does them alone.

13

Equipment

We live in a wonderful age of gadgets. There are all kinds of radios, instruments, autopilots, navigation devices, and pilot helpers. It's not difficult to spend as much on them as the price of the airplane without them.

But are all these necessary? No! What's worse than the expense is that an abundance of gadgets can often cause trouble. The trouble they cause comes from people with low experience but big bank accounts who think the gadget can do what they cannot.

During World War II I saw many Russian pilots in Nome, Alaska, pick up B-25s and other assorted aircraft we were giving them. The big question they always had about the airplanes was "Does the autopilot work, and the automatic direction finder?" We suspected they didn't know a lot about instrument flying or navigation. There was a strong radio station in Russia, and the feeling was that they put her on autopilot and followed the needle, using the autopilot to cover for lack of instrument experience and the radio station for aids or lack of navigational ability. Well, it's probably unfair to jump to such a conclusion, but it seemed obvious at the time.

However, I've known of limited-experience general-aviation pilots who do the same thing with Bonanza-type airplanes—people going on instruments "just for a little stretch" who didn't know how to fly instruments, but made it via the autopilot. Man! What a way to break your neck—not counting the illegality!

This is talking in extremes, but it illustrates the point. How much does an omni station take the place of navigation knowledge? A lot. Too many people learn the basic formulas of navigation—add westerly variation and all that—but never really practice it. They learn to tune an omni, get the radial or bearing, and fly it. They may never know where they are en route, but they are headed in the right direction.

Now, what if that radio quits, even on a nice sunny day? Do they know enough to do a good navigation job?

Last winter, when my son and I took a 402 Cessna across the Pacific to Australia, I was impressed with how little navigation I'd been using. I found myself blowing the dust from old books to bone up on techniques I hadn't used in years.

Years past, in the U.S.A., we carried DF charts to use for plotting bearings. They had radio ranges and beacons shown with azimuth overlays, a lot like an omni. When you plot radio bearings there are effects from the convergence of meridians to be applied if you want a precise bearing. I'd almost forgotten the exact method and formula. It was fun reviewing it, and important because I knew we'd be taking long-range radio bearings in the Pacific.

We don't do this any more over the U.S.A.; there are many omnis, and they're too easy to use. But I find that occasionally I may tune a radio beacon, far ahead, or one may go out of the country, on a flight, to the West Indies, South America, Alaska, or northern Canada, and there's always a chance that someday the omni might fail. Then you're in a beacon-radio, ADF bearing world. Proper plotting knowledge makes bearings more accurate and less scary. You remember that radio-beacon bearings are subject to large errors near dawn and sunset; that a long-range bearing doesn't point right at the station and you may see the ADF needle off to one side, which makes you think you're not going in the right direction when actually you may be. Or it might point dead ahead when you're not dead on. Why? Well, generally speaking, in the latitudes of the U.S.A. there's about .6° error that needs to be applied to the bearing for each degree of longitude that separates you and the station. (If the plane is west of the station, the error is added; if east of the station, it's subtracted.) That's not much if you are close in, but useful if you're far out.

You remember the techniques for using a poor bearing that swings and vacillates. You learn to watch these vacillations, note the degree through which the needle swings, and average for a reasonable center point.

You learn, too, that it's doubly important to identify a radio beacon because tuning it at night or when the sun is near the horizon is difficult on its low frequencies. It's easy to have that needle pointing at the wrong station. This can happen even on an outer marker, and it needs careful identification too.

You know to be very careful and suspicious of broadcast stations. They have to be used with caution, especially at night, when far-off stations will come in louder and clearer than a nearby one on the same frequency. To do any serious work by them, they must be identified.

Before the Pacific trip with Rob in the Cessna 402, someone told us that there was a big broadcast station in Hawaii which you could get a long way out on the California-Hawaii leg. They gave us the frequency, and, sure enough, almost 1000 miles out it was coming in strong. The needle pointed directly ahead. Actually, I thought it should have been to one side, because we were headed for Honolulu on the island of Oahu and the station was on the island of Hawaii. But, being so far out, I wasn't worried about the accuracy of the bearing. We had a long way to go, and I was sure it would begin to swing left toward the big island of Hawaii as we got closer.

But it consistently pointed ahead or a little to the right. I plotted some bearings; they didn't match as they should. Nothing bad, but not right. Well, finally we heard the station identify itself during a station break. It wasn't on Hawaii, but on the island of Maui. Then the bearings fell into place, and we were almost right on course. It wasn't important, but it did cause a slight questioning anxiety that wouldn't have been there if the station had been identified sooner. And that's one of the troubles with broadcast stations: The identifications are few and far between.

The basic navigation on that flight was dead reckoning:

Just hold a heading and wait for something to show up. In this case, something was generally a radio beacon.

It was fun to sharpen up our DR work on things like the types of courses, true, mag, compass, and a big review of the wind side of the calculator.

We made a test hop of the airplane at Wichita and very carefully swung the compass to be certain of its errors. Out there you do it by flying the section lines, which, so wonderfully, go exactly north-south and east-west. We flew them many times to check and recheck and then made up a compass card. Wichita has 9° easterly variation. How do you swing it on true lines and get a magnetic compass correction? You line up on a north-south section line and set the directional gyro for 351°—that's 000° true. Then turn, stopping each 15°, and check the heading. As you come around to an east section line, the directional gyro should read 081°. On these gyro headings you check the compass reading and see how much it varies from the gyro, and that's the deviation. Say the gyro, when lined up on a true-north section line, read 351°, but the compass read 354°. Then there's a 3° error in the compass. So to fly true north with that compass, you'd have to have it reading 354°. That makes the deviation on the heading 3° west. Why west? Because you'd have to add it to the magnetic course to get compass course; true course = 000°/360° as measured on the map; Wichita is 9° east, so you subtract from the 360° and get 351°, the magnetic course. But the compass is 3° in error and it's west, so you add the 3° to 351° and get 354° compass course, the thing to steer, without wind.

You do this a number of times to make certain the gyro hasn't precessed while you turned, and on each try smooth

out any little imperfections. Our 402's compass had a 3.5°
easterly error on the headings we'd use for the flight.

So, someone says, who cares? I never use that stuff flying
from Bloomington, Illinois, to Tulsa, Oklahoma. Well, he
probably doesn't, but someday if equipment fails and he's
suddenly DRing, it would be nice to have an accurate compass. Almost more important, he'd have intimate knowledge
of how that compass worked. It's part of the enthusiasm we
talked about, the desire to know all there is about the art of
flight. An accurate compass is required in order to have accurate radio bearings—and the radio compass should be
calibrated too.

When you DR, as we did on the Honolulu-Tarawa leg
from over Johnson Island, 1440 nautical miles to Tarawa,
and hope to pick up its beacon, you get a renewed faith in
that thing called the compass and realize it does more than
just refer to omni radials.

On that flight, early in the predawn of Honolulu, we got
the best weather information possible and carefully looked
at the winds-aloft forecast. Then the flight plan was made,
leg by leg, working out the exact drift and groundspeed,
with Rob doing the computerwork and marking it all down
as I double-checked. You're interested in your work because
those headings have to be correct. The wind forecast may
not be perfect, but you do everything as precisely as possible
and then know at least what is exact, and therefore reduce
the error to the variables that may not be exact, namely the
wind direction and velocity.

There's a feeling of being closer to the real values of flight:
the elements and man's attempt to combat them and live
with them. When you get into careful consideration of

what's going on in that sky, what north really means, and what the magnetic pole is, flying isn't just a bunch of black boxes, but it is, again, a part of man's beginning and his closeness to the earth and sky.

In the practical sense DR and holding a heading is useful on a flight of 20 miles as well as 2000. It shows us that when we're lost, holding a heading until something shows up is a lot better than heading here, then there, twisting and turning but never finding out where we are.

How long has it been since you drew a line on a map, got the winds from the weatherman, figured the courses and drift, and then took off and DRed, checking the course by map reading and correcting it as the drift varied, made estimates, and got to where you were going without once using any radio navigation? It's not only fun, it reawakens one to basics that, even if rarely used, are important. And it makes one feel, again, like an airman with helmet and goggles and the wind on your face.

The Gadgets

If I were buying a new airplane and had that big option sheet in my hand, I'd go over it with this thought: "What do I really need for the kind of flying I do?" If the gadget didn't meet these requirements, I wouldn't buy it.

Overbuying adds weight, clutters the cockpit, costs money, and is a constant maintenance annoyance because gadgets have troubles and then have to be fixed.

If we're going to be flying instruments, how much instrument flying will we do? Are we Category II folks ready to shoot a 100-foot ceiling with 1200 feet visibility? Or are we just off to fly a little IFR between places with our omni and perhaps an 800-foot approach?

For my kind of instrument flying, in a four-place general-aviation airplane, I like two transceivers, two omnis plus a glide slope, marker-beacon receiver, ADF, and transponder —and you have to have an Emergency Locator Transmitter. That's a package I can do a lot with.

It would be nice to have area navigation, Horizontal Situation Indicator, and even a flight director. But I don't need them, and I'd rather spend the extra money in getting good basic equipment—that is, high-quality transceivers, omnis, ADF, and so on. Good quality will assure me of reliability plus clout in transmitting when I want to be heard.

We haven't mentioned DME as a must. It isn't. In many ways it's a luxury. The DME approaches one is required to make are minimal and can be substituted for. ATC requests distance out at times, but can live without the information. DME is handy for groundspeed checks and estimates, but that can be done with a calculator and a watch, and makes us a more astute pilot in the process. DME is nice, and I'd like to have it, but it's a semi-luxury item nevertheless.

I'd like an exhaust temperature for getting the best mixture setting, and a carburetor-temperature gauge. These help keep that engine running and do it in an efficient way—and, of course, a heated Pitot tube.

A good stopwatch in the control wheel, as Beech does, and a clip on the wheel, as Cessna does for approach plates. I'd want a handy place for pencils and storage for the Jeps and maps.

All that in a good airplane and I'm ready to go about

anywhere. The rest is luxury, with the exception of an auto-pilot of some sort. If you're going to do a lot of instrument flying alone, then an autopilot is a very handy tool indeed, but it doesn't have to be a fancy one with all the tricky gadgets. Something to keep the wings level while you get out a chart, juggle frequencies, etc., is all. If your wife can hold it level, then you don't need the autopilot.

If I lived in the West, I'd seriously consider oxygen. High-altitude flying is more pleasant without a headache, and it's a lot safer when you're sharp and not only half with it from lack of oxygen.

Too Much

But it's so easy to load an airplane with stuff. It cuts down useful load and can either make it necessary to short fuel in order to keep under gross weight or fly with full tanks over gross, which isn't clever at all.

An airplane equipped as I've outlined can do a lot of instrument flying, but it's restricted in two areas: ice and thunderstorms. Thunderstorms can be flown so long as we are able to go around them visually. If not, then we need radar, and there hasn't been one developed yet for small, single-engine airplanes. This means we don't fly with thunderstorms in the area on instruments. Think how many times you have to stay down in a year. It's probably pretty seldom. So it's worth sitting on the ground when that kind of condition exists.

Ice: Do we need de-icers? For the props, yes; for the wings, maybe; for the windshield, yes.

Much more important is the assurance the engine will run and, if a propeller airplane, that the props are slick and ice-free. By the engine running I mean adequate carburetor heat. Most airplanes have it, but it's important to make certain yours does by test. That's why I like a temperature gauge, to know what heat rise I really have and if it's enough to keep the temperature above freezing.

Most GA airplanes are not certificated to fly in known icing conditions—which says you have to stay out of ice. But that isn't always the answer. Say we're going places and en route there's a snow-spitting, visibility-reducing cumulus deck with bases 1500 and tops 6500. No fronts, just that deck. But we want to go now, and flying underneath in the snow is unpleasant and dangerous, because we don't see very far. The logic is to climb on top and enjoy a sunshine flight. Do we get ice in the climb? Probably. Are we legal? No. Can the airplane handle it if it's a fairly shallow deck? Probably. On my Skylane I slobber that ice stuff that comes in a spray can all over the prop's leading edge and back about one third on its chord. I make sure Pitot heat is on, and then climb with fair airspeed and lots of power. I never seem to have trouble, *but* I get a good appraisal of what the tops are by asking for pilot reports and studying the synoptic situation to be certain there aren't any fronts or pressure systems within range to affect the condition. I also try to get ice reports. I mean actual reports, because the weather service will almost always paint pictures of doom in this regard.

I am not in any way suggesting one take chances, but I hate to see someone trying to fly VFR in marginal VFR conditions to stay under a cloud deck just because someone said it's icy in the clouds, when I know that four or five thousand feet above there's sunshine and blue sky.

Using the Stuff

The first step in knowing how to use equipment is to know it. Since most of the gadgets we use are electrically powered, it behooves us to know something about electricity and then, in considerable detail, the electrical system of our airplane: how it works, where the circuit breakers are, what its failure possibilities are, and how to cope with them. We should know the ampere hours available, and how many each piece of equipment uses. It's worthwhile to have a plan in mind which outlines which equipment we can get along without and how many amps we save by taking that equipment off the line—pulling its circuit breaker or turning it off.

Using equipment means knowing its limitations. We've talked about the ADF and its problems. The omni is a pretty slick thing, but it's accurate only within 2.5° (actually, 3° is more like it). And, as we all know, it's only line-of-sight. Because of that, an omni shouldn't be used until the signal is solid and the red flag firmly out of sight with the To or From indication strong.

A DME has the same line-of-sight limitation, but it is much more accurate than an omni. It's a good gadget, but lower on reliability than other instruments. It needs more shop work. Marker-beacon receivers are there, and they work, but how often do we carefully check the dots and dashes to be certain it's the marker we think it is? That's important: Just because a blue light flashes, you cannot be certain that was the outer marker. I've seen local ground radios set that blue light to flickering, and even a short time might catch a pilot's glance and fool him.

There's a manual for any piece of equipment that goes into the airplane, and the pilot should not only read it but study it until he feels he knows each thing in the airplane intimately.

Along with the mechanics of knowing the equipment and its limitations is knowing how to fly it. The basic fault is in thinking that by having equipment one can barge into very bad weather and make a successful arrival. This is the problem, and it certainly isn't the truth!

There isn't anything anywhere that will take one safely through all thunderstorms, carry all ice, shoot zero ceilings, and fly all turbulence. Since this is so, it's obvious that the pilot must be able to study the conditions he's going to fly and say, "The equipment can handle it and I can use the equipment." Or, more important, to say, "It's too tough, the equipment will not handle all that."

This is all supposing the pilot can use his equipment and is proficient in instrument flying, low approaches, navigation, and all the other facets that make up flight.

If he isn't and doesn't understand the equipment and its limitations, he's better off getting an old Cub, Champ, or something like that, and equipping it with tach, compass, clock, oil temperature and pressure. Then he should learn that equipment, how to use it, and what flying is like in its humble form before progressing to the rich and fancy. He'll be a better pilot.

14

Check Lists

The need for check lists is obvious, but there are some points to discuss.

If a check list is too long, it loses its effectiveness. People pass it by because it takes too much time. If the check list is long, then it is taking valuable pilot time from the tasks at hand, and the pilot has his head and eyes inside the airplane, not outside looking for traffic. Often the check list is read before landing, when our pilot is in the most busy traffic part of the flight and certainly when he should be looking out, not in!

The place for the long part of a check list is when the airplane is sitting on the ground before starting for the first time of the day. Then one can afford to go into considerable detail for the preflight. It really isn't a check list then, but an operational reminder.

But, when you're ready to fly, or in the air, the check list should have only the items needed for successful flight—no more, no less.

The best check list is one you make up yourself, a simple card that can be posted in a handy place. Inside the sunshade is one place. On the Skylane I have a card stuck up on the right side of the headliner where the plastic trim meets it, over the passenger's head. I need only glance that way to see it.

My Skylane check list has two parts: BEFORE TAKEOFF and BEFORE LANDING.

BEFORE TAKEOFF says:

> Fuel selector . . . . Set
> Flight controls . . . . Free (especially important
> in winter)
> Trim . . . . Set
> Doors . . . . Locked

And away we go!

Obviously we could add a long list of items, but then it isn't a check list, it's a crutch. Take the engine. We could put on the check list: mixture rich, carb heat off, prop low pitch. But are they really needed? Open the throttle with the mixture lean or the heat on and you'll know something isn't right. The same would go for having it in high pitch. Also, all these aren't likely to be set wrong, because we've just run up the engine and should have checked and set everything properly—and we should know how to run up an engine.

We might add flaps, but that's part of our thinking. In an average single the flaps aren't needed for takeoff on a normal airport. If it's short or rough or there's snow we want to come up out of fast, pilot thinking would be enough to remind us of flaps. If they are required on each takeoff, as in some twins, then we'd add it to the list, but not unless they are required.

BEFORE LANDING:

> Fuel Selector . . . . Set
> Mixture . . . . Rich

Not a big list, but we stay out of trouble with those items. Some might argue that carburetor heat should be there too, but isn't it normal to use the carburetor heat? And an important point is not to have items that aren't really put into play until some time after the check list is read. Like carburetor heat—say it's on the list. We probably read the list well before final, before carburetor heat is needed. So we set the fuel selector, push the mixture, and say to ourselves, "I'll get the carb heat when it's time." But when that time comes we've done a number of other things, are busy watching our approach, and have forgotten all about the carb heat reminder. So why have it on there?

If the airplane has a retractable gear, we put that on the list, naturally. But habit patterns we talked about previously should be set up so we remember things like the landing gear.

The point is to keep the check list simple.

What Is It?

Is the check list a reminder or a crutch? It should be a reminder. The idea is to do all the checking before looking at the list, then to look at the check list to be certain you haven't missed the important things. A pilot really should know all the items for flight and set them. The check list is to be certain that he hasn't overlooked the important points that could cause trouble. It's possible to overlook these from time to time because of fatigue, distraction, or just plain old forgetting.

If a pilot depends on the check list for everything, then he uses valuable time—valuable outside, traffic-looking time— reading the list.

Check lists started on late DC-3s as a reminder. Then check lists grew, until they became lengthy and cumbersome and we found pilots ignoring them. After realizing why pilots weren't using check lists, we made them shorter. Then they were too short, but study found there was a place for long and short check lists. Finally check lists settled down to being long where there was time to perform them, on the ground, but short in flight. Notice the following 747 check list. We'll take the BEFORE LANDING portion; it's divided into two sections: LANDING PRELIMINARY and LANDING FINAL.

The LANDING PRELIMINARY is read when the airplane is still out of the traffic area and above 18,000 feet. The items:

1. Seat-belt sign
2. Anti-ice

3. Logo light
4. Altimeters
5. Gross-weight and airspeed bugs (a tick that's settable on the airspeed indicator for approach or reference speed on landing)
6. Go-around EPR (a power to set throttles in case of go-around)
7. Cabin altitude (flight engineer item)
8. Circuit breakers (ditto, in case any were pulled in flight)

Notice that some of these items are lengthy, like checking the gross weight and referring to a chart for the approach airspeed. That's why they're done when the airplane is out of the airport traffic area, 18,000 feet and 15 minutes before landing.

The LANDING FINAL goes like this:

1. Spoilers (they are armed for landing, something the pilot can do by feel without looking at them)
2. Brake pressure (a quick glance at a gauge)
3. Altimeters (an extra check on the proper setting)
4. "No Smoking" sign (quick flip of a switch)
5. Ignition (push four buttons to be sure it's on for jet landing)
6. Fuel heat valves (done by engineer)
7. Fuel boost pumps (ditto)
8. Cross-feed valves (ditto)
9. No. 1 ADP (ditto; an air-driven pump for the hydraulics)

After the landing gear is extended there are three items:

1. Gear and anti-skid (to be sure it's down and anti-skid is on, a quick thing to check)
2. Flaps (quick to check)
3. Altimeter setting on international flights (because you come off 29:92 at transition levels which are low, like 3000 feet, and you might miss it)

So that's it to get that big complicated airplane on the ground safely. A simple, quick check list that has been made up after deep study by the best in the business. Since it's simple, each item is important, and pilots are expected to pay attention to each with complete seriousness, and they do.

A check list isn't something to read without registering. If we read "Mixture," we should look at the mixture, have it sink into our mind, check it, and then go on to the next thing. For that tiny split second our mind is all mixture and not thinking about something else. That's the importance of check lists: to remind us and draw attention. That's why they should be as short as possible, with important, serious items. When check lists are like that, we'll use them—which, after all, is what we want to do.

A silly thing I've seen people do is memorize the check list and then recite it without looking at it. That destroys the entire idea. Because there's a tendency to memorize the list after a while, it's not a bad idea to make up new ones from time to time, rearranging the items so you'll have to look and notice when the list is used.

New airplanes are coming out with annunciator panels

and various types of "automatic" check lists. These are fine, but they add the problem of being certain that the electric panel is working properly, plus that it seems easier to gloss over this type of presentation and a pilot must make himself pay attention even more than with a printed list.

But any list is better than none.

15

The Case for the Glider

I fly gliders for fun and research, and because it's the most fascinating flying there is . . . to me.

I'm not trying to sell gliding, or soaring, as they call it. I don't manufacture gliders, distribute or sell 'em. I don't own stock in a glider factory, and actually it's to my personal, selfish advantage if the world has fewer gliders; then I won't have to wait for tows on busy days or fly against so many hotshots in contests.

But I've found soaring makes good pilots, pilots who understand weather and the techniques of flight better than those who don't soar. So I pass along these points despite my selfish desire to have most of the soaring sky to myself.

Soaring means learning to fly on the ragged edge of stall,

because much of glider flying is turning in thermals to climb, and you want to be on the very slow side to keep the turn radius small and be at best speed for maximum lift, which isn't far above stall. So flight near the stall becomes routine.

My son learned to fly gliders at age fifteen. When he was sixteen he started power flying in a tail-dragger Cessna 120, and I didn't teach him myself, because I think good instruction is done by a specialist and just because I fly it doesn't mean I can instruct. So I found a very capable young man, Jim Frankenfield, to do the job. One day I asked him how Rob was doing.

"He sure can handle slow flight like I've never seen before."

"Have you ever instructed anyone who was a glider pilot before he took up power?"

"No, I don't think I have."

"Well, Rob isn't any super pilot. It's simply the fact he's spent a lot of time on the edge of a stall in a glider and, like most glider pilots, isn't afraid of the slow end."

And it's a fact. I can almost tell if an airplane is being landed by a glider pilot. I see it all the time at Warren-Sugarbush, where there's lots of glider flying. When the glider pilots arrive in power ships, they use minimum runway length and land with grace. But sometimes you see an airplane landing with too much speed, touching down halfway up the field, burning it on the ground. In the snack shop, where pilots gather for talk and to watch airplanes, there'll be a chorus of "Who's that?" because it couldn't be anyone who flies gliders that we know.

That isn't to say that power pilots who don't fly gliders

can't fly. There are many who do it beautifully, without the benefit of glider experience, but the power pilot who has flown gliders has a certain edge over most non-glider pilots.

Because one doesn't have power, each landing becomes a precision thing with the approach carefully adjusted to touch at some pre-chosen point. It adds a dimension of exactness to flying.

In DC-3 days I was an assistant to the chief pilot at La Guardia, and one of my jobs was to give new copilots transition. One of the copilots was Loren Petry, or Pete. (Now he's an international 747 captain.) I knew that Pete had flown gliders almost from childhood.

"For the heck of it, Pete," I said, "I'd like to see you do a three-sixty power-off spot landing in this ole DC-3."

He did it with ease. We tried some one-eighties, and other power-off landings. They were perfect—and he hadn't had any previous DC-3 time! I tried it with non-glider-flying copilots, and they didn't do nearly as well.

But the more interesting point was that when I asked the non-glider copilots to do this, I could see an incredulous look pass over their faces along with a slight fear of flunking the maneuver until I told them it was only an experiment and wouldn't count. But when I asked glider pilot Pete, he lit up, laughed, and thought it would be fun. Power-off flying wasn't a mysterious world to him. The others did well after a few tries.

It was interesting. How far we've come—you cannot mess around and have fun like that today with an airline airplane. I did have something like it, though, when I took my rating ride in a Cessna Citation. The last thing the inspector had me do was make a spot landing.

We picked the spot he wanted on the Wichita airport, one of the painted distance marks on the runway. It was a snap to put it exactly on that spot. And this isn't in any way supposed to be an immodest statement, but just to show that, one, the Citation is easy to fly, and, two, flying gliders made the entire maneuver very routine. I've also done it power-off with 747s on line trips; the 747 has a glide ratio of about 18 to 1.

Glider pilots become very conscious of wind and turbulence on the approach and near the ground. They watch for sink off the end of the runway, the times extra airspeed is needed, and when a lifting effect may make the landing long if not corrected promptly. Once a person has flown gliders, this part of flying has become automatic; it is ingrained, a part of his nature.

The landing area is important because in today's flying one sees too much evidence that people aren't being taught to land airplanes, but rather to "drive" them on the ground with little finesse.

The system seems to be to have lots of speed, because it's supposed to be safe, and then plunk it on the ground, all three wheels at once—the three wheels being those of the tricycle landing gear, which has had a lot to do with this kind of flying. The tricycle gear is a big safety plus—fewer ground loops, better taxiing visibility, fewer nose-overs, etc. —and people tout it as one of the greatest things that ever hit flying.

Well, that may be, but the tricycle gear also has created a few problems. What it did was allow landing without the gentle touch. It means most any clod can drive an airplane on the ground. It means too-fast approaches that end up off

the runway, and overshoots with too-late power applica-
tions, and pileups on the far end.

Some instructors, bless them, teach people finesse despite
the tricycle gear, but some don't, and when pilots get on
their own, especially after being impressed by loudmouths
who brag, "I always carry extra speed, I ain't a-gonna spin
'er in," they fall into sloppy habits.

And don't think this is just the newish pilot. I've seen the
disease on airlines. A pilot looks up the reference speed on
an airliner; it's 30% above stall for the weight he's using. It's
a good number and was decided upon after long tests and
years of experience. Specific cushions are added when the
surface winds are strong and gusty. But I've seen airline
pilots arbitrarily add 10 knots or more as a routine matter.
Periodically an airline will have to check and instruct some
pilots to get them back to proper speeds as readouts on
recorders show up excessive approach speeds . . . or the
occasional dramatic demonstration by a slide off the runway's end.

This is an area, again, where one realizes that the pilot is
there to take care of the unusual, to know when extra speed
is really required and when the approach must be slow, and
when it can be normal. That's part of the ART of flying.
How beautiful to watch an approach at proper speed, the
end of the runway cleared with enough room but not too
much, and then, because the speed was correct, a gentle
flare and smooth landing! It's as good to look at as a lovely
painting.

The $1.3V_s$, 30% above stall, is ample under normal conditions. When it's gusty and windy we all carry a little extra to

fit the situation, but that can be overdone and you see a pilot come in with a great excess of speed, try to flare, zoom 50 feet, go back at the ground, zoom again, porpoise, and finally slam it on. "God," someone says, "what a wind!" Well, probably 75% of those wild antics weren't the wind but the guy flying it much too fast.

One can learn this area well if he learns to fly a tail-dragger, too, which, if I were king, everyone would have to fly at one time or other—preferably early in instruction.

But this is only part of the things glider flying teaches. It makes us better mountain pilots. Much glider flying is done on ridges and in waves, which gives an appreciation of what the wind can do when mountains get in its way and what this can in turn do to the airplane. It does good things as well as bad.

Warren Ketcham, one of our Sugarbush gang, flew to Nevada with wife, Mae, in their 125-horsepower Cub. It was an interesting trip, but over the western high country he struggled to climb the mountains. Being a glider pilot, he hunted out some wave action and got up and over by using its rising air.

Knowledge of waves keeps one out of severe turbulence where the rotor lies in wait downwind of the mountain, ready to really fling one about the sky.

Consciousness of wind against a slope will make an aware pilot fly the upwind side of the ridge, getting lift which makes for extra airspeed. The unknowing could be on the downwind side, struggling to keep altitude, and losing speed.

Thermals are useful, if one is familiar with them, in helping to climb with a marginally powered airplane. A cloud street is recognized and a pilot may move his flight path over a few miles to get under a cloud street and go much faster in the lift.

The sky is full of these things, and a glider pilot becomes intimate with them and uses them in power flying. I've used waves to help climb a heavy cargo 707 from Milan when I wanted to cross the Alps and go to New York.

Taking off from Honolulu in the Cessna 402, 30% overloaded and headed for Tarawa, the airplane got into the air pretty well, but then wouldn't climb. It didn't take us long to realize that the mountains to our left, with the wind from that direction, were making that entire area one of settling air. We made a turn as soon as possible to get over the sea and away from the hills. After a few miles we flew out of the sinking air and the airplane got back to its normal climb rate.

The air we fly in is constantly moving, not only horizontally but up and down as well. A complete pilot is conscious of these movements, and glider flying makes him alert to the air's motions.

Airplanes are often misjudged, called poor climbers and slow cruisers by pilots after one flight. Our quick-to-judge pilots might not know that the poor flying could have been caused by the air mass motion and not much at all by the airplane.

Air moves on a big scale, and many times I've seen this in jets as I climbed, heavily loaded. I used to count on it going out of JFK for Europe and adjust ATC altitudes accordingly. If there was a new high pressure in the area, I knew I couldn't make as high an altitude by Nantucket omni as I could if climbing in a low. Why? The front of a high is a big mass of colder air which is sinking. Basically the air mass is settling, and you don't climb as well because you're climbing "uphill." Conversely, a low is warmer, with the air converging and lifting, so you climb better. It's a subtle difference, but real when you're struggling for altitude.

Where we do our soaring in Vermont, the airport is located in a mountain setup that invites all sorts of subtle air movements. They are fascinating to study in a glider, and each day, each weather condition, makes these air movements somewhat different.

The west side of the valley has a north-south range of mountains that go up to 4000 feet. The valley is about five miles across, and on the east side there's another north-south line of mountains, almost 3000 feet high. The bigger mountains to the west cause waves, and we have them to some degree almost any day. Certain days the wave is weak and you can barely fly higher than the mountains, but you can find lift—perhaps only 100 or 50 feet per minute more than the glider's sink rate, but it's enough to stay in the air and fly all day if you want to. Waves like this, you find, tend to be of long length, so that the air over the entire valley is softly rising.

Other days, with more wind and a new air mass, the wave may be strong and steep. You tow toward the big mountain and go through a rotor that bounces you around like a cork. It sometimes seems crazy to be on the end of a rope hooked to the tow plane ahead, which bounces, settles, zooms, and then you do the same thing in a moment. Both of you are below the mountain and headed right for it. The smart maneuver would seem to be a one-eighty and getting away from the cul-de-sac against that mountain. The bare ski runs before you show the rocks and roughness.

Then, just when this wild, crashing turbulence is at a peak, it stops, stops dramatically, and what was wild, rough air suddenly is the smoothest you've ever flown in. That's the wave, and you pull the release. The tow plane goes down and left, back to the field; you make a tiny right turn and then head toward the mountain. The smoothness of the air is startling, and you look around at the trees, mountain, and sky, wondering what happened. Your eyes go to the instruments, and the variometer's needle points to 600 or 800 feet per minute up! The altimeter hand is winding around, and in a moment you are even with the top of the mountain, then above it, and Lake Champlain comes into view, with New York's Adirondack mountains lumped up behind, a vast, majestic view. Oxygen goes on as you pass 12,000 feet. Before the rate of climb hits zero, you may be over 20,000 feet!

Generally these days are clear days, and in the fall the view is magnificent, with trees in full color, the strong mountains, peaceful valleys, and white New England towns below. It's a joyous time.

Sometimes the wave has a lenticular cloud. Lenticulars are special clouds because they are lens-shaped, as they follow the wave's pattern, smooth on top, and don't drift away; they stay with the wave. They stretch long and narrow the length of the valley, and you cruise above them at will.

Flying an afternoon in waves, climbing, descending, drifting back to get in the secondary wave, even the third, downstream, watching their effects, the differences in sink or lift in relation to the mountains, and the cloud, is fascinating and instructive. But, even more, it creates an intimacy with the sky and its movements that takes away fear as it builds understanding. After you land at the end of a day like that, you may be in the company of friends, busy with the ways of the ground, but your thoughts are with the wave and sky. You are detached, like someone with a new love. You're different from that day on.

Each stage of glider flying brings new wonders and awakenings, new enthusiasm, but it's when one starts cross-country that the ultimate is approached. It is surpassed only by flying competition, which is racing cross-country.

The sense of accomplishment in going from one place to another entirely by your wits is almost staggering. There isn't any way to cheat or be helped. You start from the friendly home airport into an unknown adventure, staying up, advancing, covering distance.

The skills are finding lift and climbing, which is reading weather and constantly observing subtle changes: An overcast ahead will cut off heating and lift, so you go off course where there isn't an overcast and hope to work back to your destination later. You watch thunderstorms, stay away from the blowoff overcast part where there's sink, and try to nudge along the front edge where there's lift, but staying far enough out not to get in it. You watch for lenticulars, or telltale signs of a wave, study a ridge for the wind blowing against it so you can soar its length and gain miles. And these are only a part of the observations required.

You fly in concert with the ability of the sailplane. You know its performance at all speeds and how to use it for the conditions at hand. You speed up in sink, slow in lift, and have a bag full of tricks that apply to your sailplane.

You navigate, not by omnis, ADFs, and all that, but by reading a map with intimate detail. Try making a dozen or more turns in a thermal and then come out of it on the right heading and oriented to the way you want to go. It requires a certain skill. Map elevations mean more; the topographic lines are studied in detail so you know which way a ridge turns and twists and how the wind will hit it and give or take away lift.

During all this there's the thought "Where will I land if I cannot find lift?" And while all the other thinking is going on, there's that conscious thought toward the possibility of landing. It's important and necessary, but you learn not to let it be the consuming thought of your flight. So you develop an ability to keep important things important but do other things as well, and do them very well.

Most gliders are towed aloft behind an airplane, and that teaches a part of flying we don't experience any other way. It's an interesting trick to stay in position or correct for out of position when it's turbulent and the tow plane is all over the sky. It creates flying delicacy, plus ability to observe and react, which, although you don't spend your flying life being towed by another airplane, does make for more confidence in flight and improved flying ability.

The importance of towing came to me in Grenchen, Switzerland, when I wanted to rent a glider and fly over the beautiful Jura mountains. They required a checkout with a check pilot in an old two-place glider, which was fine by me.

We towed behind a Cub, and it was the wildest tow I'd

ever been through. Normally tow planes bank gently, but this one did steep turns, reversed direction, zoomed, leveled off, even descended for a brief time. "Gad," I thought, "this idiot can't fly! What's he doing towing?"

I learned later that this was part of the checkout, to see if I could hold position in tow under extreme conditions.

Such are the things glider flying is made of. If I didn't fly gliders, I would be missing a big part of the wonders of flight.

16

Rules and the FAA

Once upon a time there weren't any rules for flying. There wasn't any government bureau bothering people aloft. Then, a little before Lindbergh's flight to Paris in 1927, Washington got to work licensing pilots and airplanes. Old-timers tell me the government had quite a time making barnstormers and their ilk take it seriously; they wouldn't get licenses or license their airplanes, because they flew intra-state and felt the Federal Government didn't have jurisdiction except when airplanes crossed state lines. To some it was kind of a joke anyway, and they ignored the rules interstate too.

The Department of Commerce, Aeronautics Branch, was the government bureau. They'd put a red tag on an unlicensed airplane to signify that it was grounded, but the pilots would tear it up and go on flying anyway.

The story goes that, just before Lindbergh took off for Paris, someone discovered that he didn't have a pilot's license. A Washington bureaucrat thought it would look bad, in case he made it, not to have him a U.S.A.-licensed airman, so they sent him a license!

But that was long ago, in the good old days . . . at least in that sense. When I went to work for TWA in 1937, Air Traffic Control was only in a few major areas; the rest of the time we were on our own. The pilots kept tabs on each other, and when you made an approach into Columbus you knew where that American Airlines flight was. Airlines schedules now are so big and voluminous that you can't know them all, but then you knew every schedule that could possibly conflict with yours. In most cases you knew the pilots, too, and, by cooperation via radio contact, kept from occupying the same air space at the same time. We did pretty well, too.

The idea today of pilots keeping their own separation makes the FAA go into deep shock, because they have some odd notion that pilots aren't responsible-enough people to be allowed this authority. Well, pilots might not be able to maintain an orderly flow of traffic into a busy terminal, but they certainly could in the less busy areas of our big country. Giving them the authority, along with devices technology can produce to make it possible, would make flying more useful and flexible, plus saving taxpayer money and staving off user charges.

But that's another argument and cause. What we have, we

have, and the FAA, with its Federal Air Regulations and the ATC system, is a fact of life.

Pilots have to deal with these facts almost as much as they do with the law of gravity, but always the law of gravity comes first! A pilot must not be intimidated by the FARs and allow them to affect his best judgment.

This is an important and delicate point, because the FARs and all that go with them are so complex that it is difficult, maybe even impossible, to fly any flight so that somewhere along the way a regulation isn't violated.

This fact shouldn't, of course, make us think we can toss rules out the window and go our merry way. There are good regulations, many good ones.

The FAA Gives

And the FAA gives us a chance to break rules when it's really needed. This important relaxation is called "Emergency Authority." It's in the same part 91.4 I talked about in "Pilot in Command." I quoted paragraph (a) of 91.4; now here's paragraph (b):

> In an emergency requiring immediate action, the pilot in command may deviate from any rule of this subpart or of subpart B to the extent required to meet that emergency.

"This subpart" is A, General Rules; subpart B is Flight Rules General.

This sounds great, and is another part of the FARs we shouldn't allow them to remove or dilute.

They Take Away

But there's a paragraph (c) right behind it, and that says:

> Each pilot in command who deviates from a rule under paragraph (b) of this section shall, upon the request of the Administrator, send a written report of that deviation to the Administrator.

Now, we have to talk about that, because it's loaded!

First notice with great care the little part that says, "upon the request of the Administrator." Here we play the old Army game: Don't write anything unless you have to! Next, if you have to, write as little as possible. Keep it simple, and write it with the knowledge that probably it will get into the hands of FAA's legal staff and some obscure thing may be twisted into a rope with which to hang you! Write it with your lawyer, and don't let him get fancy either.

I've been through this once personally, and many times when I represented pilots for ALPA. It can be a tough road. State facts and think ahead about how they could be used against you. I'm not saying you should lie about anything. But the truth should be simply presented.

Let's say we get a load of ice and ask ATC for a descent clearance. They say we can't come down because of traffic. The ice is getting tough, airspeed going down, and we have difficulty maintaining altitude. So we declare an emergency, turn off course, descend, and tell ATC.

FAA asks for a letter. We write one and say:

> Due to severe icing it was necessary to declare an emergency.
>
> > Sincerely,

Simple, but it opens a can of worms. What were you doing there in the first place? Why did you wait so long to do something about it? Did you get adequate weather briefing? Is your airplane certificated for operation in known icing conditions? Was there other action you could have taken that wouldn't have required an emergency?

As you write the original letter, you'd better be thinking about answers to such questions. But—and this is important—don't pre-guess their questions and attempt to hedge against them in your original letter. Maybe they'll never ask them—but if you bring them up originally, they'll be gone into for sure. If they come back with questions after your first, simple letter, you'll know exactly what they are thinking and what you have to answer. It's not a nice way of doing business, but that's the way it is. The rule: Don't say anything unless requested, and then say as little as possible.

Which, though not exactly in the same category, reminds me of an incident when the 707s arrived on the scene and we went to school. Our ground school was directed at passing the FAA oral. The first pilot to take it failed, because he overtalked. The inspector asked about a certain hydraulic valve and its off-on action. Charlie, the pilot, answered the question properly, but kept on talking about the hydraulic system and got himself all wound up and confused. The inspector told him to go back, study some more, and retake the exam.

I was next and, having heard the story, wasn't about to say anything. The first question the inspector asked was "Do you know the limiting speeds?"

My answer: "Yup." There was a pause as I sat waiting for the next question.

"Well, what are they?"

I told him a few.

"How about the hydraulic system?"

"Yup." Pause.

"Well, I guess I'll have to be specific."

"Yup."

We both laughed about it, but I stuck to "Yup" as much as possible.

The fact that a letter is required following the use of emergency authority makes people afraid to use the authority. They shouldn't be. This letter is necessary. And the FAA in most cases acts very reasonably when emergency authority is used.

There are times when emergency authority is used wrongly, not by the man in the air but by the man on the ground, to intimidate and, as I call it, blackmail pilots.

I've heard it more than once from ATC. Maybe you are holding and it's turbulent. You call ATC and request to leave the holding pattern, change altitude, or whatever for relief. He says, "No." You bounce some more, then tell him you've just got to have another altitude. He comes back:

"Are you declaring an emergency?"

It's a method to shut me up that absolutely infuriates me.

It works on the timid soul in the pilot, and possibly may get him in trouble because he's afraid to declare an emergency if it's really needed. In such conditions the pilot is attempting, by good flying practice, to get out of a condition before it does become an emergency.

Fortunately, most ATC controllers are cooperative and try to help work with your problem, but there are the occasional short-tempered, uptight few who get tough and use this form of blackmail. My answer to them is something like:

"No, I'm not declaring an emergency, but if I don't get relief I may have to!" This flips it back to him a little, because you're on tape as warning him and in a hearing you could make a lot of points with that conversation.

Emergency authority is a good thing; it cannot be done much differently than it is. The final outcome depends on the FAA's being reasonable in its investigation of a pilot taking the action, and in the reasonableness of pilots' using it only when really needed. *But no pilot should ever be timid about using emergency authority if he feels it's necessary— FAA, letters, and all!*

There Are Good Guys

An organization as big as FAA with its thousands of employees is bound to have some bad ones, but, at least in my experience, the good ones far exceed the bad. Most all dealings I've had with FAA inspectors have been pleasant and helpful. I think of a General Aviation District Office (GADO) like the one in Allentown, Pennsylvania, headed by John Doster. He's been around the business as long as the rest of us; he knows the failings and solid reliability of pilots. He understands that operators have to make money. You can talk to John or his staff and they'll go all the way helping to solve your problems. They can be tough when they have to be, but they're tough with as velvet a glove as possible. John isn't the only one; there are many like him and his gang.

And a Few Bad Guys

Unfortunately, all of the people administering the rules aren't sweet and compassionate. I was at a small air show when the FAA arrived in the area. The traffic was sort of mixed up, but it was working okay. The FAA inspector-pilot didn't like it, and said so on unicom, where all could hear:

"Clear the airport! This is the FAA. I'm running this show, and if the field isn't cleared I'll cancel the show!"

The ground announced to all that the FAA wanted to land, so clear the area. Of course, thirty-seven private aircraft arrived with little or no fuss before and after him.

This guy, getting excited, was quick to show that he was Mr. Authority and we'd better all bend down and kiss his foot. The FAA lost a lot of ground that day.

Down on the Potomac

There's no doubt there are two FAAs: the one in Washington and the other out in the field where the action is. Often the ones in the field, like John Doster, live in spite of Washington, where the morass of bureaucracy slows and befuddles action. In Washington rest bureaucrats who are more sensitive to political requirements than those of flight; in Washington rest FAA types who are steeped in tradition, superstition, and the old way of doing it; and in Washington rest some good guys who stand in a constant bloody state as they forge into battle anew each day, trying to make it better and get the job done.

Above it all, like an ominous cloud, is the FAA legal department. Blame them when you attempt to crawl through the complex regulations and rules, because they wrote them with all the protective, cagey, and sneaky language lawyers write for each other, not for you the pilot!

To prove it, just get in any airport bull session on some part of the FARs. There will be as many interpretations as there are pilots in the room. Look, too, at the thickness of the regulations, how many there are.

I was on loan once to the FAA for a brief period to do a report on landing accidents for the Administrator. I had to have a spot to work from, and the only one they could find was a cubbyhole down in the legal department. I sat with these types in coffee breaks and an occasional lunch. The experience scared me. So often I heard arguments concerning violations, and most times it was law-school argument on how the law read, not a whit about the merits of the case, the violator's background, or bowing to a suggestion for clemency from an inspector from the field.

The classic one, back a number of years ago, was the airline pilot who was almost knocked out of the sky by an airplane which cut across his nose. The offending airplane's pilots were practicing under the hood, in the Washington terminal area. The airline pilot got on the horn and made a loud complaint. The result: He was violated by the FAA because he had admitted breaking the rule which says you're not to fly closer than 500 feet to another aircraft! That's when the airline pilots went underground and stopped reporting near misses. The FAA finally gave immunity for those reporting near misses, because they weren't getting any reports. This was revoked recently, and again near-miss reports aren't being made by pilots, which hides our traffic danger picture.

The most scary legal thing to me are weather minimums: what you can and cannot land with. Get a group of airline pilots in school during recurrent training and observe the class on regulations. There's little agreement. Here are the pros, and yet on the tricky ones like local surface conditions, sliding scales, crosswinds, turns to parallel runways, and a bunch of others, you'll get almost as many answers as there are pilots.

One night at JFK a skyful of airliners all landed successfully, yet shortly after the FAA filed violations on the pilots because they'd landed below minimums! They didn't think they had, but a sneaky point about downwind component made the landings illegal.

What do we do about all this? Suffer, unless you have the magic formula for changing bureaucracy—something that's needed in a lot of places besides the FAA. We shouldn't give up trying.

As pilots we need to know the regulations and rules as well as possible and, where we don't know them, hope that good practice and common sense will see us through. One of the delightful things about flying internationally, which I did for some twenty-seven years, was that I wasn't under the constant eye of the FAA. Now, of course, the foreign countries have caught on and have their complicated regulations too; generally they're similar to the FARs, but perhaps a bit more tolerant.

The FAA, regulations, and rules, too, from states and communities, are with us, big and real, but in dealing with the people who administer and enforce them we can feel, generally, that they are good folks who will give an even break or more. That's been my experience with FAA people in the field and some in Washington—but if I have an argument with the FAA I want to keep it in the field if at all possible.

Now and then we'll run across the bully tough guy, and that's bad. There aren't many, thank heaven, and there isn't anything that says a citizen cannot get tough right back at him!

But the main theme is that we are pilots in command, charged with successfully flying aircraft. That book of rules, and the horde of people behind it, should never intimidate or distract us, or make us do something we don't think is in the best interest of flight.

17

The Airplane's Environment: Cold and Heat

Airplanes have the ability to transport us from one type of environment to another before the mind and body have a chance to get in tune with the difference. Even at 100 mph we can fly from cold New York to warm Florida in a day if you get up early and fly late; it was a routine thing in the '30s.

Airplanes fly from dry desert to wet tropics; low altitude, rich with oxygen, to heights that will not support life for more than moments; we can take off in shirt sleeves, hot and sweaty, but shortly be freezing; we are able to leave civilized areas and be over inhospitable, uninhabited country in minutes; the airplane crosses oceans few ships ply.

A good airman has a feel for these differences in climate;

his knowledge causes him to guard against their pitfalls automatically. He is able to anticipate them. It's the kind of instinct that makes a Down East lobsterman squint into the sky at mare's-tail clouds as the clammy east wind alerts him to foul weather coming. These are the things experience is made of.

The chances of ever being forced down are slim; most pilots will go their entire flying lives and never land except on an airport or, at worst, a pasture on some farmer's field.

But because airplanes take us over such inhospitable country, that long-shot forced landing may be a tragedy. The warm comfort of a working airplane cabin very quickly can become a cold, lonely place far from civilization.

To think you will take emergency equipment if you go on a long trip isn't the way it works. You need to think about it on every flight. There is inaccessible, tough country very close to civilization. Sweep the land north, east, and southeast of the busy Los Angeles area: You can be lost and unfindable within a 30-minute flight! Cut the corner of a body of water like one of the Great Lakes and you're over a sea; the interior of Florida, not far from Miami, is a tropical hell; New York State, almost within sight of Albany, is wild forest. I helped look for three days before we found an airliner that had crashed within 30 miles of Utica, New York. Many years ago a Cub took off from Louisville, Kentucky, with an hour's fuel supply, and it has never been found.

Even though the chance of being in such dramatic surroundings is slim, it can be lifesaving to realize that an airplane is different from anything else, with the possibility of being a pampered modern person in plush comfort one moment and Robinson Crusoe the next.

It certainly isn't necessary to dwell on this fact, but any sensible flying person will carry an emergency kit, as we have discussed, and read carefully a good book on survival. I've always liked, and go back to for review, *The Survival Book*, by Nesbit, Pond, and Allen (Van Nostrand publishers). Such information should be part of a pilot's education.

These changes in our environment are things we deal with to some degree every time we fly.

COLD

Cold relentlessly fights the airman. It is tough!

When it's running, the airplane flies beautifully. It gets off quickly in the dense air, climbs fast, feels great. That's the good part. But we have to get it started before we fly.

The first important point is: Be prepared for cold, and realize that you cannot do things as quickly as on moderate days. Being prepared begins at skin level, with long underwear. During World War II you had to clear into the Alaskan theater through Great Falls, Montana. There they carefully checked to see that you and the airplane were properly equipped for the cold world north. Long underwear was a requirement, and you had to undress enough, right in the operations office, to prove you were wearing longjohns before they'd allow takeoff for Fairbanks and points north.

Covering for your body is a matter of having enough, but not so much as to become overheated.

To me hands, feet, and ears are the big items, and I can freeze even at 20°F if they aren't warm. Mittens, though awkward around an airplane, are the only things that keep my hands warm. I wear some kind of hat that has ear protection. For feet the name of the game is having them covered, with the covering loose enough not to stop circulation. In Vermont the winter footgarb worn by all the old-timers—and I watch them to see what I should wear—is the leather-topped rubber-bottomed felt-lined boot. That combination will take tender feet through almost any temperature. They are a development of the Eskimo mukluk, but more practical.

Even if it is cold where you start from, it's such a temptation to dress lightly when you're headed south, expecting to be in warm weather within hours, wondering what to do with the excess clothing. That's true, and the stuff is bothersome, but if it takes a while getting going you can become awfully cold. Worse, if you have that rare forced landing while still in cold country, silk socks and loafers are going to be mighty silly, and cold!

If it's very cold, the airplane should be heated in advance. At Montpelier, where I hangar the Skylane, Edmando Roberti, proprietor of Vermont Flying Service and old-timer in these parts, who really knows his stuff, has all the heaters, equipment, and know-how necessary. I call in advance, and when I get to the field the Skylane's nose is wrapped in canvas and a heater is pouring in warmth. Starting's a cinch.

Some folks keep a hot lightbulb inside the engine cowl all the time while the airplane sits in the hangar. This scares me a little from a fire viewpoint, as it sits alone and untended during the long nights. But it helps.

The instrument panel needs preheat as well as the engine. Gyros get stiff and instruments sluggish from the cold, and a quick takeoff into instrument conditions can be hairy with a slow responding horizon.

We put five heaters in the B-17 in Alaska, one for each engine and then a long tube of heat up into the cockpit area—not just to make it a nice warm place when we climbed in, but to get all the gadgets working properly. Not only in cold places, but at all times, it should be an automatic part of taxiing to check the directional gyros and turn indicator as you wind your way to the runway, to be certain they are up to speed and working.

Starting Cold

Frank D. Comerford, President of Comerford Flight School, Inc., Hanscom Field, Bedford, Massachusetts, runs an excellent school and knows a lot about cold weather operation. With his permission, I reprint below part of his Winter Operations bulletin. It's good, and I've used the ideas successfully.

1. Plan your flight like any other flight, but remember that sunset is earlier; and unless you're night-qualified, you have to be on the ground earlier.

2. When cleaning off aircraft, use brushes or brooms; if ice is on the surface, do not try to break or shatter it—you may damage the metal skin on the aircraft—use heated hangar or other device.

3. Do not use windshield scrapers on aircraft—they will scratch windows & paint.

4. Check the static port and gas tank vent lines for ice—breathing on them helps.

5. If the key won't go into lock, breathe on lock; pre-heat key to melt ice in lock.

6. If when turning on the master switch or starter switch, a click is heard but nothing else works, it means the particular solenoid is frozen and will not actuate; heating the offender will usually clear the trouble.

7. Don't take off with snow or ice on the aircraft, and don't plan on the snow blowing off when you start your take-off; the snow may blow off, but there can be a layer of ice under the snow, and the aircraft will not fly. Frost on the wings will prevent flight!

8. If the aircraft will not move when you attempt to taxi out, it may be that the brakes are frozen or the tires are frozen to the ground. Rocking up and down on the wing-tip will sometimes free this condition, but heat is better.

9. Avoid taxiing through puddles as the water splashes up into the brakes and eventually freezes. This is very important with retractable landing gear airplanes as the gear may freeze in the up position and be impossible to extend.

10. When taxiing in snow, follow in other wheel tracks and apply back pressure to the elevators to lighten the load on the nose wheel.

11. When taking off from snow-covered runways or fields, use the soft-field technique.

12. Smooth application of the throttle on take-off, a go-around, or a touch-and-go landing is essential. Abrupt use of the throttle tends to flood the engine, causing it to cough, hesitate, or even quit completely.

13. During cold weather, plan approaches with partial power so as to prevent the engine from over-cooling. Using partial flaps for drag is a good technique.

14. Avoid landing on ice-covered landing strips with a crosswind; but if you must land, make a full-flap landing on the up-wind side of the runway, and do not use brakes during the roll-out.

15. Treat snow showers like rain showers. Fly around them or away from them and you won't get caught on instruments.

A Discussion of Cold-Weather Engine Starting

Before attempting to start an aircraft engine in cold weather, there are a few points that the pilot should be made aware of:

1. Aircraft batteries, because of weight limitations, are small and therefore have less energy, by far, than an automobile battery.

2. Any battery has less energy when it is cold than when it is warm.

3. Aircraft electric starters are smaller, because of weight limitations, and therefore have less force or drive than automobile starters.

4. A cold engine is much harder to turn over than a warm one because the oil is thicker—also, cold metal contracts so that bearings, etc., are tighter and offer more resistance to turning.

5. Any engine, cold or warm, that has been allowed to stand for a long period of time (12 hours or more) develops an oil seal which makes it difficult to turn over at the start until the oil seal is broken.

6. Engines which have been shut down by mixture control cutoff are dry of fuel and require many revolutions to draw fuel and air mixture into the cylinders before they can possibly ignite and start.

7. Cold engines that start and run for just a short time (30 seconds or less) usually develop frosted-over spark plugs and won't start without preheat.

Putting all this information together, it's easy to understand that one doesn't have too much time or battery energy to fool with when starting a cold aircraft engine in cold weather. Therefore, a method must be devised to minimize the failure rate of cold starts.

Nothing can be done to change items #1, #3, and #4. To correct #2 would require keeping the battery in a warm place and then installing it in the aircraft just before attempting to start—and this is not too practical. Items #5 and #6 can be taken care of by pulling the engine through manually for several revolutions with the mixture control rich and with a few shots of prime prior to manually turning the engine. (Caution: be sure the magnetos are off.) Item #7 can be avoided by taking the necessary steps to prevent the engine from quitting once it starts. After all this, we will still have only a precious few moments of turning before the battery will give out, and we will have to use them wisely.

This brings us to preparing the induction system for an immediate start as soon as the engine is turned over. During the time that we are pulling the engine through by hand to break the oil seal, we could also be pulling a fuel charge into the various cylinders, and this would save the battery from having to do the same work. We accomplish this by charging or priming the cylinders—by priming with the primer first (mixture control rich), pumping the throttle a couple of times, and then closing the throttle.*

When we are pulling the engine through, we are not only breaking the oil seal but are also charging the cylinders with fuel. An important point here is that as we pull the engine through by hand, we should be listening for two things:

---

* Each pump of the throttle squirts about a teaspoonful of fuel up into the throat of the carburetor, and pumping should not be overdone because of carburetor fire hazard. Closing the throttle after this is for the purpose of restricting the air flow through the carburetor—too much air and a cold engine won't start. The throttle should never be above a medium-idle setting.

1. A clicking of the impulse couplers in the mags as the engine is pulled through compression. No clicks . . . no start. Get a mechanic.

2. The air being sucked in by the carburetor should have a *wet* hiss to it. No wet hiss indicates that the carburetor is not drawing fuel. So, no wet hiss . . . no start.

Pull the engine through about six times or six blades of the propeller. Now the engine is *almost* ready for starting—just a few more preparations.

Enter the cockpit and pump the throttle two or three times to prime or charge the intake manifold, and then close the throttle and crack slightly—not above medium-idle position. The throttle must stay here for the remainder of the starting procedure and until the engine is running steadily. Open the primer and pull out—and leave it out. Now, and only now, turn the mags on, master on, and energize the starter. The engine should start almost immediately, within one or two blades of the propeller.

Start pushing on the primer to keep the engine running, and as soon as the primer is all the way in, pull it out again and be ready for when the engine starts to die down. As this happens, start pushing the primer in slowly to keep the engine running, and repeat for as long as it is necessary.

In the meantime, apply full carburetor heat to start heating the carburetor. Notice that all this time nothing has been done to the throttle. To apply more throttle during this early start period would only kill the engine with too much air.

When the engine is finally running smoothly without priming, lock the primer and shut off the carb heat. You're now ready to taxi.

If the engine should start initially and run for only a short period (less than 30 seconds) and stop, then the spark plugs will be frosted over, and the only way the engine can be started then is by preheating of the engine or of the spark plugs.

Easy Does It

An airplane on the ground in winter is a difficult-to-control vehicle. The traction between the wheels and icy taxi strips and runways is poor. The plane becomes a weathervane, and the wind wants to push it around. There isn't any situation that makes you realize more that an airplane belongs in the air!

There's one major answer to handling an airplane on ice. The old Italian expression fits perfectly: *"Piano piano* [Softly softly]." Take it easy and go slowly.

Applying brakes by the off-and-on method will give surprisingly good braking. The cycle of off-and-on will be a second apart or less. As everyone knows, if brakes are held the wheel will skid, but as the brake is first put on, before it skids, it does a little grabbing, even on slick ice. So the trick is to put the brake on for that split second of time that it holds, then release it—and then put it on again for another little grab. This off-on action is exactly the way anti-skids work on big, fancy airplanes, except that intelligence-gathering gadgets tell the brakes when to go off and on, which they do automatically. They make a big difference in stopping distance—about 50%, and in some cases more. The poor man's way is to do it yourself.

I flew our B-17 a lot in the Aleutian Islands, and there one finds the nasty combination of icy runways and high wind. A more perfect weathervane than the B-17 was never invented—with its big fin and rudder, it wanted to weathercock in a breeze! We had some exciting moments on the icy ground trying to stay within revetments and cross bridges that went over ditches and streams. I vividly remember approaching one of these bridges on Attu in a strong wind, the surface covered with ice. I pussyfooted along, braking on and off. Just in front of the bridge a gust of wind hit that big fin, and I saw we were going to make half bridge and half ditch as the B-17 started to weathercock! My flight engineer, Barney Dowd, was standing between the copilot and me. He was intent on our progress and the copilot, Garth Sharp, was all eyes on the action too. Unfortunately, I didn't have time to tell them what I was doing, but as we started to weathercock I saw that the only possible way out was to spin around in front of the bridge, get stopped, regroup, and try again. I blasted two engines on the side opposite the wind and increased the weathercocking action. We spun around in front of the bridge without hitting it, and came to a stop headed into the wind.

"Man!" shouted Dowd. "You scared me to death!"

"Sorry, old boy, but I didn't have time to tell you what was going on."

Surprisingly, though, the off-on brake method got us around some very icy places.

Nose Wheels

A nose wheel can fool you on ice, because there isn't much weight on it so it doesn't have much traction. It'll go sideways just as easily as straight when it's slippery. Making a running takeoff—that is, to taxi into a runway from the taxi strip at an angle while pouring on the coal, expecting to get straight with the rudder as you accelerate—is a sure way to end up off the runway when it's slippery. The nose wheel doesn't grab, and there isn't yet enough airspeed for rudder action, and all of a sudden there's that awful out-of-control feeling, with you just going along for the ride.

One icy day in Paris I gave a copilot the takeoff. The taxi strips and runways weren't very bad, but they did have some slick stuff. To my surprise, he pulled into the runway with the 747 and, without getting lined up, poured on the coal and went right into the takeoff run. That big ship started going cockeyed as the nose wheel just skipped sideways across the rough ice and snow. I knew what would happen, so I took over before, I'd guess, we had 40 knots, but even at that slow speed we almost went off the runway, and I had one of those "Now I've got it, now I don't" experiences that scare you silly. I breathed a great sigh of relief when we got stopped still on the pavement.

On slick runways, as I told that young man, you gingerly taxi into position, get the nose wheel straight, stop a moment to be certain all's set, and then open 'er up for takeoff. This also is a time to be very conscious of where the wind is coming from. If it's across the runway, you can plan on some interesting action during the early part of the takeoff run, and had better be prepared for it.

Landing has a lot of the same problems after you get on the ground. Fortunately, when it's cold we land slower, and every advantage should be taken of that. It's the time for a short approach, because there may not be any brakes.

One cold dawn I was landing at Orly. Everything was routine as we descended through the foggy, cold mists and picked up their excellent runway lights. Just as we flared, the tower called and said, "Braking action nil!" The runway was a slick glaze of ice. We touched down and immediately used reversers, which are good at high speed but don't help much below 90 knots. The 747 also had to be out of reverse below that speed, because if it wasn't, the engines would bang in compressor stall, which in turn could damage the engine about $250,000 worth! (That was an early 747; they're better now.) After 90 knots we were dependent on brakes, but even with the anti-skid our rate of slowing was pathetic. It's a terrible feeling to see the far end of the runway come closer, especially as you wonder why in hell you didn't pour on the coal and leave when the tower gave that late information. But it was a split-second decision, alternate airports weren't good, and somehow, with false ego, I thought I could get her stopped. The anti-skid worked hard, but we slowed slowly. I was trying to think what lights and obstructions we'd take out when we slid off the runway, how much damage I'd do to the airplane, and what I'd say at the hearing! But, like a movie thriller, we came to a stop not 100 feet from the runway's end. I taxied in carefully, feeling a little shaky but mostly feeling stupid, and lucky. I did a lot of thinking on that bus ride into town.

Cold Aloft

At very cold temperatures engine gauges tell different stories from the ones we're used to, and they may be a little scary.

I left Fairbanks, Alaska, for Pt. Barrow one very cold winter morning, 50°F below. The four engines ran smoothly, but the head temperatures had us worried, as they stayed at 90°C, less than half their normal temperature. We nervously waited for the engines to sputter and die, but they hummed on beautifully. Back home I asked the Wright Company about this: How low a head temperature would she run with? Essentially the answer was that if the engines were running, the head temperature was enough no matter what it said. On the low side, of course.

Oil Is Different

The oil temperature does crazy things when it gets cold. Sometimes it goes up, way up. Generally that means a frozen, congealed oil radiator. Sticking out in the breeze as it does, the oil doesn't get warm enough to flow through the radiator, so there isn't any oil cooling, and less circulation. It gets hot. The cure for this, if you have a radiator control, is to pull it to the hot position—that is, call for even warmer temperature, crazy as it sounds. That will heat the oil, free the congealed radiator, and bring the temperature back to normal.

All this won't happen, however, if the airplane is properly

prepared for very cold weather with a winterization kit. And that's what an airplane should have.

The times we find ourselves in very cold areas with a warm-area airplane, perhaps for one quick winter trip north, the answer is to get the engine warm before starting. Take the time to preheat so all the oil and radiator are warm. Often this saves broken radiators, because if the engine is started with thick oil in the radiator, the pressure-relief valve may not work and the surge of pressure will burst the radiator. The B-17, in its early days, was plagued with this because its relief valve wasn't very good. Barney Dowd and Bill Foley, our two flight engineers, changed more than one radiator under awful conditions in the far north. A modification finally came that cured the problem; it was cause for wild celebration.

Time Alone Will Do It

Everyone tells about not taking off with frost on wings, and it's true. It is also one of the reasons cold-weather flying takes time and requires getting up one to five hours earlier than we would in sunny, tropic climes. We need the time to get the airplane organized.

To prevent wing frost problems, some use wing covers, but I've had lots of trouble with them. They are fine if it stays cold, well below freezing. But if the ice melts and then refreezes, the covers will stick fast to the wings, and getting them unstuck is a difficult, cussing job. I'd rather just swab her down with alcohol.

Inside Too

That cockpit should be warm before starting, as I've said —not for comfort or the instruments alone, but also to be certain that the windshield will be clear of frost, with nice warm air flowing over it. DC-2s had a flexible hot-air tube about one and a half inches in diameter. We'd move it around and point it where we wanted a hot spot on the windshield. (Sometimes we'd stick it down our shirt fronts to get warm!) It was pretty good, but covered only a small area, and a lot of flying was done looking through a space about six inches in diameter. Now and then the hot-water boiler, which supplied heat, froze and there wasn't any hot air. That's why pilots carried a putty knife in their kits: to scrape the frost off the windshield—and often the ice on the outside, too. That was a job of opening the side window and reaching around in the cold blast to scrape clear a hole to see for landing.

If an airplane doesn't have adequate airflow on the inside of the windshield to keep frost off, a piece of transparent plastic you can stick inside the windshield or side window will make an airspace between plastic and window that will keep it clear of frost. Kits for this are found in auto-supply stores in northern places. We use them for winter flying in gliders, which don't have any heat. Fortunately, even the most humble modern airplane has a system that will keep the windows and windshield clear—if it's warmed up enough before takeoff.

Warming Up

Paradoxically, while we want engines warm, we don't want to run them too long on the ground. The modern airplane has the engine cowled very tightly, and unless it has good airflow engine hot spots develop that will harm it. It's very important to run up into the wind, but to run up as little as possible. In very cold weather this, again, means lots of preheating to be certain that the oil is flowing. In mediumly cold weather, about 20°F or so, the engine, most manuals say, is ready to go as soon as it's warm enough to take power without blurping or faltering. The individual manual should be read to check this out. But the important item is to not sit on the ground, as we could with old radials, and run and run them. They are made to cool and work properly in the air, and that's where they belong.

Landing Cold

In low temperatures we have to keep engines warm so fuel will vaporize. We have to keep it cozy, and for certain this makes winter landing approaches power approaches. Sometimes, in very cold—and that's defined as less than 14°F—it means power approaches with quite a lot of power. These approaches aren't bad, because engine reliability at the reduced power setting should be excellent. The engine isn't being asked for high power output or strain, just enough to keep warm.

Considering that the runway may be slick and we want to land as near the end as possible, and that we need power during the approach, the landing will require a little tricky work. If the air is calm, as it often is in winter, we can make a nose-up, power-drag approach and then chop it just as we clear the "fence." If it's gusty and windy we cannot risk this, but the wind itself will help land short and stop sooner, so we can still use reasonable airspeed and power to get on the runway's near end.

Other Things Aloft

If it's very cold, the altimeter isn't telling us how high we actually are. We're skimming Alaskan mountains at 10,000 feet; the outside air temperature is —40°C. Our actual altitude is a little under 8700 feet. Even —20° puts us down to 9400 feet. The lower altitude can be subtle, and even when it's only somewhat cold the altitude difference can be serious when we're trying to fly marginal VFR in poor terrain. An experienced pilot automatically gives himself a little more altitude when it's cold, and if it's very cold the computer comes out to find out just exactly what the difference is.

A low approach in very low temperatures makes for altimeter errors and, of course, in the wrong direction. At a "warm" —20°C., (—4°F), the altimeter error will be about 25 feet when shooting a 200-foot minimum. That may not seem like much but, combined with other possible errors, it adds up, and certainly is a good case for not cheating and going lower than the law allows.

Visibility

Visibilities in cold weather are wonderful, and the sight of a clean horizon 50 miles ahead is exciting, but when snow falls visibility can suddenly be no better than in fog. The most difficult aspect of snow is the white-out, with the sky and ground blending into one flat white mass. I talked about it in sensory illusions. It's an instrument-flying world, and trying to sneak over white ground, contact, in a snowy sky, is close to impossible.

It Isn't All Bad

Cold flying is a matter of preparation and equipment. It takes work and care, but the rewards of crisp performance and clean air with inspiring visibility are often worth the bother.

Arctic weather isn't as bad as we think. There's much less weather than in temperate areas. Severely cold air cannot hold much moisture, so it seldom snows in the far north in the middle of winter. There's little or no ice that forms on the airplane, and most precipitation is hard ice crystals that don't hurt anything. To make bad weather we need temperature and moisture contrasts. These don't come together until we get to the southern part of the Arctic. Even in the temperate areas, like the northern United States, when there's a severe outbreak of cold air aircraft icing isn't a problem, and generally the weather is excellent. When warmer air begins to creep up from the south and the temperatures moderate, then it's time to look for weather trouble.

No, cold isn't bad. If you don't mind wearing some extra clothes and being a little inconvenienced, cold can be very nice.

HEAT

Sunny climes certainly make flying life much easier. It's a quick start and off we go. We pay for this warm convenience with reduced takeoff and climb performance, as most any student pilot knows. But perhaps he doesn't know that high humidity, which comes with the high temperature of the tropics, makes performance suffer. The Air Corps, operating B-29s out of Guam during World War II, took humidity into consideration. Air with lots of water vapor is lighter than dry air, so it's less dense and we don't perform as well.

Surprisingly, the cool times of hot climates are sources of trouble. One is carburetor ice. With high humidity, and the refrigerator action of the carburetor, carburetor heat is more of a necessity than in very cold weather. We aren't relieved from using carburetor heat just because the weather is hot.

In dry desert country carburetor heat isn't that important, because it takes moisture in the air to make ice and there isn't much over deserts, but it's worth watching in case there is some humidity from a nearby sea or other body of water.

Fill 'Er Up

The idea of filling tanks before putting the airplane away for the night is especially good in warm climates, where large temperature changes occur during the day-night cycle. These changes can cause condensation in the tanks. Full tanks, of course, leave little room for water formation. And this calls for careful draining of sumps before takeoff in hot climates.

Weather

We don't see as well where it's warm. The nuclei of salt particles, sand, and other contaminants form haze. VFR flying isn't so easy, even on nice days.

In the tropics, over the sea, the visibility isn't as bad as the central United States in summer, but cloud bases are lower. On a hot day scattered cumulus in Kansas may have bases of 8000 feet, but at an equal temperature in Miami the bases will be less than 3000 feet. There's more water vapor in Florida, and so the condensation level is lower. Farther south, near the equator, the cu bases will be even lower.

Airplanes run well in the heat, but care is needed to prevent long run-ups and overheating.

Brakes cannot always be parked, because the heat will expand the hydraulic pressure and possibly burst a line, although most modern airplanes have relief for this.

It's excellent to have some ventilation in the cabin when

it's parked, because a sealed cabin sitting in the hot sun can reach impressively high temperatures, which are hard on equipment—like the "pretzelization" of a plastic computer.

The Pilot Too

The environment of heat makes for fatigue, and perspiration depletes us of salts and liquids. After takeoff on a hot day when we've sweated getting ready to fly, we've noticed how up where it's cool we relax and pretty soon are fighting to stay awake. Part of that is fatigue from body salt and liquid losses.

We can get nasty burns from hot cowlings. I remember my first indoctrination when I was flying an open cockpit, Pitcairn Mailwing, to Los Angeles. The engine got rough, and I landed at Palm Springs to look it over. The J-5 engine's magnetos were on the front, with little pieces of metal cowl over them. I landed, got my tools out, and then went up to take the cowl off the mags and check them out. I couldn't handle the cowls; it was like picking up a hot frying pan.

Hot Sun Beats Down

Now we're inside, and the sun doesn't beat on us and things like lip balm and sunburn lotion aren't as needed as they were in open cockpits. In gliders, however, sitting under a canopy, we wear tennis hats to protect our face and neck from sunburn, as well as use lip balm.

When I was a kid flying an open-cockpit biplane, my lips gave me pure hell as the sun burned them to a crisp.

Sunglasses

I've always been a firm believer in the idea that we overdo the use of sunglasses. The more we use them, the more we need them. An eye surgeon told me to use them when my eyes really hurt and needed protection from glare. That's pretty much what I've done, and though they're in my flight kit, they go on only when I'm flying into the sun, or close to the top of a brilliant cloud deck—something of that nature. I guess I haven't been far wrong, because after 40-plus years flying I need glasses only for reading in dim light. That's very unscientific, and maybe the luck of having good grandparents. But I've talked with people about this and from a corner-drugstore sort of survey find that pilots who use sunglasses sparingly seem to have better eyes, or eyes that stand up better.

Hot-Climate Areas

Hot-weather flying, on the weather side, is generally a matter of thunderstorms, and they are either air mass, frontal, or orographic. Air mass means to fly early in the day, because they form later from heating. Frontal is a matter of studying weather; orographic means to stay out of mountains when conditions are ripe for thunderstorms.

Over deserts there are sandstorms, and a good sandstorm can reduce visibility to zero. Knowing if they're coming is a matter of studying the weather forecasts. If a sandstorm is predicted it's nothing to take with light heart. I've seen sand kick up in Egypt to the extent that you might as well try to land in a dense fog. It's very tough on the airplane, too, unless you want it sandblasted.

Desert areas can have a reduced visibility condition when the wind is nil or light. Fine particles of sand that float in the first few hundred feet of the atmosphere do it. I've made many landings at night in Cairo, Egypt, where the airport is out from town in the desert. As I crossed the coast near Alexandria, the sky would be crystal clear, stars where you could touch them and the small clusters of lights from towns in the Nile Delta visible miles and miles ahead. But on descent, from about 500 feet on down, there would be a slight haze from, I guess, the fine silicone particles in the air. The lack of contrast because of the sandy desert floor below made it very difficult to tell where the ground was. As when flying in snow over snow-covered ground, it was necessary to keep a careful check on the altimeter to be certain where the ground was. There are sand hills south of Cairo airport, and there was a rash of accidents on those hills as airliners flew into them on approach. The hills were difficult to see in the no-contrast, hazy air. They have finally put lights on these hills, but this reduced-visibility-landing in desert sand is worth attention.

The Tropics

Generally the tropics have good weather, except for the hurricane season. But far enough south, near the equator, one finally meets the intertropical front, or, more fancily, the Intertropical Convergence Zone—the area where the northeast trade winds bump into the southeast trade winds and shove the air up to where it will condense into showers and thunderstorms. This area wanders north and south of the equatorial region with the seasons: south in winter and north in summer. If you go from one hemisphere to the other, you'll fly through it.

During World War II it was a constant problem to the new, inexperienced pilots. Early in the war airplanes were routed to South America and across the South Atlantic. The pilots had to traverse the front, which would be somewhere north or south of the mouth of the Amazon.

This front can look awful. The clouds are black, yellow, green, and other colors depending on how the sun hits them. They build to tremendous heights. Any self-respecting thunderstorm builds up to the stratosphere; that inversion shuts them off, and that's where the tops are. As our meteorological books tell us, the stratosphere is highest at the equator and lowest at the poles. It'll go 60,000 feet or more in equatorial regions. So the thunderstorms build that high, and such heights are impressive.

We lost a lot of young men because they'd see these scary cloud masses and try to duck around them. Often there wasn't any way around and they'd get lost trying, forgetting navigation as their interest in staying out of the clutches of the wild-looking clouds became all-consuming. We'd never hear from them again. Sad.

These thunderstorms can be rough, but they don't approach the turbulence and wildness of a good old Kansas thunderstorm. There aren't the temperature contrasts, and while a tropical storm will bounce you, it's impressive mostly for the amount of rain. And how it does rain! Sometimes you think you're in a submarine instead of an airplane, and wonder how the engines ever keep running in such a deluge. But they do, especially if you keep a ready eye toward carburetor ice and use heat carefully.

It's Not Always the Same

This intertropical front's intensity will fluctuate from day to day, and some days one can wander through the clouds and have a very pleasant ride. On others, it's solid, wall to wall and black.

Taking the 402 to Australia, we encountered it between Guadalcanal and Brisbane. We'd had showers north of the 'Canal, but they were easily duckable. South, Rob, who was flying, ducked shower after shower, keeping track of time off course and all that for navigation reasons. But finally it was all black and no way around. We didn't have radar, so we couldn't pick out cells.

"What do we do now?" Rob said.

"Just pull down on your belt a little and we'll bore right through that so and so." I knew we'd bounce and get very wet, but I also knew it wasn't going to be like a front between Omaha and Kansas City. It was a big letdown, as we didn't hit much turbulence. The showers weren't bad; it was a weakish day. But we bumped in and out for a few hours, and finally slid out from under the upper cloud deck.

"Nuts," Rob remarked, "I thought it would be tougher than that." He was disappointed.

Actually, I've always enjoyed flying the tropics and the Intertropical Convergence Zone. There's a zest to that kind of flying.

Tropical flying is interesting, and especially appealing now, as I write this, because it's snowing outside, and places like Tarawa, the 'Canal, and Captain Cook's Islands seem very appealing.

18

The Airplane's Environment: Mountains and Sea

Up high airplanes and people don't breathe well, and it's cold. The coldness doesn't always mean things act cold. Sounds backward, but it isn't. I'm always impressed, when flying a reciprocating engine at high altitudes, how the outside air temperature goes down but the head temperatures go up! At $-50°C$ I've seen head temperatures approaching their upper limits. Why? Because there isn't air with substance flowing over the engine to carry away the heat even though the outside air temperature is very low. There just aren't enough molecules.

And this really is the whole story of high flight. The figure that impresses me is that half the earth's atmosphere is in the first 18,000 feet . . . give or take a little.

This high world, where it's cold and there isn't much in

the way of the gases our bodies need for life, is a spooky place indeed. There's an overpowering loneliness at times, and a feeling that security is far away. But, like all other problems of flight, preparation and knowledge make it hospitable.

First is that, no matter how big and strong we are, we cannot get along without oxygen. So long as we recognize that it's necessary, we might as well use it properly. This means great respect for the equipment: clean masks checked properly and carefully taken care of when not in use. It's pretty silly to toss unused masks into a corner, where they can get dirty and damaged. Masks are the direct connection with life and deserve fitting care. So do hoses and the oxygen bottles. The entire system isn't kid stuff, it's for real.

As most anyone knows, oxygen is dangerous stuff. Lots of airplanes have been blown up during oxygen-filling sessions. We know that dirt, especially grease, and oxygen together mean fire. And it's important to know, as Sailor Davis taught me, that when turning off that big valve on the bottle one shouldn't give it an extra twist to be certain it's closed, because this can "bite" off little pieces of metal in the valve and make filings that, with oxygen, can catch fire. He taught me that oxygen bottles are turned on and off with slow, careful turns, and never seated down hard.

While on oxygen, there should be a constant check to be certain it's flowing and getting to the crew member. If there's more than one crew member, they should have a procedure to check each other. Flying alone, we need to know all the time that the regulator is working and we're okay. I look at my fingernails and check their color; if they seem bluish, I do something quickly to get more oxygen.

During DC-4 days I had an experience that enlightened me to the possibility of crew members not being checked enough. Although normally the DC-4 didn't fly high, I had one at 18,000 feet one night, trying to get on top so the navigator could obtain some star sights and find out where the devil we were.

We were on oxygen. The navigator, who had to stand on a stool for his head to be in the astrodome, had a long hose to supply his oxygen mask. We didn't pay much attention to him, because we were interested in staying on top and flying the airplane so it would have as little motion as possible to make his shots better. Suddenly there was a hell of a crash: It was the navigator falling off the stool, out cold. The stool was on the oxygen tube and had pinched it off completely. He not only didn't have enough oxygen, he didn't have any! (We revived him with no damage done.)

In the B-17 every crew member was on interphone when we were on oxygen, and the cockpit made periodic calls to each one for confirmation that he was alive and well.

Happily, today most high-altitude flying is done in pressurized airplanes, and we aren't even aware of the need for oxygen. It's a pleasant shirt-sleeve world. Of course, pressure can be lost, and then it's a difficult world.

Actual cases of a sudden pressure loss are very few when the amount of flying in a pressurized environment is considered: thousands of hours every day. But because of this we shouldn't be complacent about it. The flying crew has to have oxygen loss as a big item in their handy bag of "what ifs." I get into pressurized general-aviation airplanes and very seldom see the pilot check his oxygen masks as part of the preflight. Airlines have it on the BEFORE START part of the check list. Often on GA airplanes I see the mask jammed down in a corner with junk all around it, and I wonder how long it's been since the pilot simulated a decompression and practiced putting the mask on fast.

The FAA rules for crew members' wearing masks at very high altitudes shouldn't be taken lightly, because loss of pressure at altitudes some GA airplanes fly—40,000 feet and above—is practically instant death without oxygen.

Twice I've gone through the Air Force oxygen indoctrination with lectures and a session in a chamber. It was fun, and very useful. These sessions can be arranged at many Air Force bases, and are well worth the bother.

The last one I did was at Edwards Air Force Base before they'd allow me aloft in an F-104. That session included an explosive decompression. They had me at a low altitude and then in an instant zoomed the pressure to 35,000 feet by puncturing a diaphragm. It wasn't as bad as I'd thought it would be, but it was wild! Air escaped through all orifices of the body, foggy condensation filled the chamber, and I was startled for that split-second moment, and sat wondering what was going on and what to do about it. Even knowing what was coming, I had that "goof-off," do-nothing pause that comes with emergencies. How much longer would the "pause" have been if it had happened unexpectedly?

The chances for catastrophic effects from explosive decompression in a big airplane are much less than in a small one. A 747 could lose a window, and maybe the person sitting next to it, but the pressurization system would keep sufficient pressure to give ample time for descent to a safe altitude before the rest of the passengers were out cold or blithering idiots.

But a small six- or eight-place airplane is different, and a window loss would be very difficult to live through in the time it takes to put on the oxygen mask and get it working at very high altitude. Wearing a mask, by one crew member, is extra important in these smaller aircraft.

Wings Need Air Too

There are certain techniques in flying the airplane up high related to the thinness of the air. Simply, the airplane isn't flying well. It's clawing to hang on, and abrupt, big movements of the controls can result in a mush, or stall, with big altitude loss before flying under control again.

The most dramatic case of this is someone trying to sneak over the top of thunderstorms in a mushing, near-stall condition. The turbulence, which extends into the clear air above the storm, can easily dump him, and wild control movements to correct only make it worse.

But high-altitude poor flying can extend much lower, because it's related to the airplane. A 747 is mushing and struggling at 41,000 feet, but a 90-horsepower Cub may be doing the same thing at 10,000 feet. The Cub can be struggling at that altitude and yet not be very high above the terrain as it tries to grunt its way over mountains. Clumsy flying could dump the airplane right into the hard stuff.

Altitude effects extend to the ground as we attempt to take off from high-elevation airports. The experienced pilot is always conscious of airport elevation and the terrain elevation around the airport; it's one thing to get off the ground, but another to clear high terrain if it is close on course. One of flying's creepy feelings is to be on instruments, climbing poorly, and wondering just how close the ground is. Sometimes it's advisable to climb in a holding pattern over a fixed point before proceeding on course. Instrument departure procedures don't always reflect the poor flying qualities of some airplanes at high altitudes on hot days. This is each pilot's responsibility: to consider his own situation and fly to fit it. ATC instructions, vectors, and paper procedures notwithstanding, THE FINAL RESPONSI-BILITY FOR HAVING ENOUGH ALTITUDE BELONGS TO THE PILOT—ON ARRIVAL, DEPARTURE AND AT ALL TIMES!

When it's hot at altitude, as we all know, the density altitude is higher, and even a lowish airport at 2000 feet can be impressively high when the temperature is up; at 100°F it is really 5000 feet! And we can find 100° lots of times in summer.

High-altitude effects are obvious, and a good pilot understands them. But an important point to remember is that altitude is relative, and we don't need big numbers, like cruising at 45,000 feet or taking off from a 9500-foot airport, to make us think altitude. Altitude relates to the performance of the airplane and the heat of the day. Sometimes 1000 feet can be high. As pilots we are never relieved from thinking altitude effects when we fly.

MOUNTAINS

Sometimes I've felt guilty flying over mountains. I look down on peaks like the Matterhorn, from 1000 feet above it, and realize the difficulty a climber has in reaching the summit while I sit in comfort and, without effort, enjoy its grandeur.

In a glider I skim along ridges, only 100 feet above them, looking at trees, rocks, and small washes where ravines are born and drainage starts toward the valley. I fly across a col and the earth drops away with breathtaking suddenness. Then, in a moment, the next ridge is reached, and I fly close to it feeling that I'm looking at places men don't see; just the birds and animals and I know this wilderness.

Mountains are exciting. Just when the boredom of the long westbound flight of flat sameness across Kansas seems overpowering, one suddenly notices the lumpy shape of the Rocky Mountains beginning to rise out of the far horizon, breaking the flat vista. The flight is exciting again.

Flying over mountains is part of the wonder and joy of flight, but it makes certain demands and requires care and knowledge.

Mountains collect weather, cause turbulence, offer limited landing places, and present the problems of flight at high altitude.

As we know, weather is caused by lifting air, and mountains in the path of air lift it so that sometimes weather is formed even when the general weather is good. On a summer afternoon those Rocky Mountains will be covered with huge thunderstorms while the plains to the east have only fluffy cumulus.

Wintertime snow and ice hug the mountain, and it requires high altitude to clear the tops. Trying to stay underneath to fly up valleys and through passes is difficult and dangerous. Flying a pass requires good visibility and gentle wind velocities.

On still, clear nights cool air drains down the sides of mountains, and the valleys will be closed with dense fog until the sun has been up long enough to burn the fog off.

Wind and Mountains

Wind increases its velocity and causes turbulence when it flows over mountains. The curve of the mountain acts like a venturi and looks a lot like that picture we see in aeronautics books showing how the air speeds up over the top of the wing, reduces pressure, and makes lift. That happens on mountains, and Mt. Washington in New Hampshire, only 6288 feet high, has recorded winds of 231 miles per hour! Now, it's obvious that winds at 6000 feet don't reach 200 miles per hour. But the shape of the mountain causes the winds to move faster than the ambient winds. This can happen to some degree on any mountain. Imagine the spillage and turbulence on the downward side of that!

In mountain areas the pilot is conscious all the time of the wind direction and velocity. With that knowledge, he stays away from the downwind side, the bad side, because that's where downdrafts and turbulence make the air nasty. In gliders we feel pretty frisky on the upwind side of a ridge and will run very close to it, but we're very, very careful not to cross over the peak and get on the downwind side. It could dump us right into the ground.

If there's a valley, pilots favor the side where the wind is starting up the slopes. If the valley is narrow, it may have turbulence from the downwind side over most of the valley and even up the lower slopes of the mountain across the valley, with wind blowing up it. In narrow valleys, with high winds, we want altitude above everything, and the stronger the wind the higher we should be.

Downwind of ranges we find waves. If clouds are forming we may see telltale lenticulars that signal a wave. Then it's time to be careful of the rotor which is under the wave, generally at about the height of the mountain. It will extend higher than the mountain, but not be as violent there as lower down—but violent enough.

In a way the wave rotor is another facet of not flying close to mountains on the downwind side. If waves are suspected, then the farther downwind from the mountain the better.

Waves form on ranges, but if there's a wide cut in the range or it bends around out of the prevailing wind, there probably will not be waves in that area. But narrow cuts can increase wave turbulence because of vortex action . . . so be careful! Approaching a range of mountains, flying into a head wind, we can suspect waves and turbulence. It's time to study our map carefully, and look at the range through the windshield. If we can find a pass going through the range, or a place where it turns a corner out of the wind, then that's the way to go: The wave will be less, or nil. We'd fly high at any rate, but height doesn't always keep us out of rotor turbulence, and in the West, where mountains are high, you can have a fairly wild ride in the 30,000-foot range.

Stay High

In mountain areas we may be operating at high altitude where the airplane's performance is marginal. Then it's important to stay as far downwind of the mountain as possible and have enough altitude to give working room if we mush or, in an extreme case, stall out. If the airplane is struggling to stay up, or climb, then it isn't smart to be close to the terrain.

If We Have to Land

Forced landings can be tough, because fields are few and small in high terrain; we'll be landing fast, and use more room than at sea level.

The valleys are where the fields are, and we want to clear ridges with enough altitude to get into the valley ahead or behind if she quits. Sometimes valleys are narrow and the choice of field pretty poor. We con ourselves into thinking we can get under those wires, next to that bend in the stream, and get her stopped. But suppose it's hot summer and that field has an elevation of 5800 feet?

The general answer is that the poorer the field availability, the higher we fly, and if we cannot fly high then we detour to follow valleys and better terrain. West out of Albuquerque, the Santa Fe Railroad wanders generally north of the direct course, and often, in a single-engine airplane, I've followed it so I could have a relaxed flight.

Twin-engine airplanes aren't always the answer, because most light twins are marginal on one engine anywhere, and pathetic at altitude, and an engine loss only means a slower descent than in a single unless it's turbo supercharged. Many of them shouldn't be flown with a philosophy much different from that for a single.

That High Wind Again

Our Mt. Washington case of the 231-mph wind: This high wind creates a lower pressure and false altitude reading near the mountain, lower than in the general area. That's one reason minimum instrument altitudes are 2000 feet over mountains, rather than the 1000 feet for flat terrain. So if winds are strong, it's clever to have extra altitude for this pressure loss. And, as always, it's very important to have the latest altimeter setting for the nearest station, and especially so in mountain areas.

Love 'Em

But flying mountain areas is fascinating. There's never-ending diversification, mystery, and beauty. If I couldn't fly over mountains, I'd be missing much of the joy of flying.

SEA

Our flight paths will eventually take us near or over the sea. It is scary, mysterious, fascinating, rewarding, and beautiful. Understanding its endless distance is like trying to picture eternity.

My ocean flying has been in land planes, and I relate all knowledge and preparation to that kind.

The first time over the sea was at age seventeen in a Pitcairn, single-engine biplane, Miami to Havana. On the map it looked like a little stretch of water. I had the vague idea you'd see land ahead or behind almost all the way.

I flew down the Keys, circled Key West once, and then picked up the heading. In a few moments the Keys disappeared, and nothing loomed ahead except the blue sea. It was a shock! The loneliness of the endless sea engulfed me; the thought of a forced landing was terrifying. I even considered turning around.

But I held the heading, and within fifty minutes the dark outline of Cuba came into view. For the first time I felt the exhilaration of making landfall, the joy of escaping from the uncertainty of the sea, of being back over friendly earth, of having navigated well.

Many years and thousands of crossings later, the sea is more friendly, and being far from shore doesn't seem lonesome, but landfall is still an exciting time.

There's one thing important about flying over the sea: There aren't any mountains to run into, unless one comes up to a mountain island when on instruments and too low. But that's an outside chance, to be sure. En route it's safer than land. Statistics show international air travel safer than domestic.

The sea demands accurate navigation, sufficient fuel, and emergency equipment.

Navigation can be anything from holding a heading until

the radio beacon comes into range to having the ease and sophistication of an inertial system.

This navigation demands accuracy, careful flight planning, good compass corrections, which we talked about, and precise accounting of time. The navigation is interesting and exciting. It connects old sailing ships with airplanes and gives the pilot a sense of being close to basics that entitle him to be called navigator as well as airman.

Navigating over the sea, distances are generally long enough so that the comfort and ease of omnis and DMEs are out of range. Radio beacons are less accurate. The pilot needs to know all the forms of navigation, and a complete study of a good navigation book, like Lyon's *Practical Air Navigation*, is necessary—and interesting. Real navigation is used over the sea. Our normal airways navigation is actually only a knowledge of gadgets, how to read and set them up. Deciding which way to turn to intercept a radial, the way to enter a holding pattern, or bracket an ILS is very far removed from the Polynesian who understood substellar points and used them to cross thousands of miles of ocean and find small islands in the Pacific. Learning navigation that will be useful to a pilot over the sea relates him to that Polynesian navigator and man's real knowledge of dealing with this globe; he learns the skills that are the art of navigation.

Be Prepared

The sea will seem more friendly if there's emergency equipment to use if you find yourself on that sea. You can build an overwater-emergency kit of any shape and size, but some items are essential: life vests, to start—and they should be worn when overwater, at least in a single-engine, and not tossed back in the cabin somewhere. For another, a raft with a cover so a canopy can be erected to keep off the brutal sun, and catch rainwater. There should be water and either a desalting kit or a solar still. The *Survival Book* will tell all the needs, and then you can fit them into your operation.

Almost as important as the raft and water is a signal device. The easiest is a signal mirror, then flares, and finally an emergency radio. But if I could have only one thing, it would be that signal mirror. You can do wonders with it.

Fishing tackle, food, and all the rest are best chosen after the compromises of weight, space, and the area you're flying over have been determined. If it's cold northern water, you'd need an immersion suit. Let's not kid ourselves: Being wet at low temperatures is about an impossible situation. It's going to be tough to make it, and probably you will not.

In warm climates it's tempting to fly in short-sleeve shirts, but I always wear long sleeves because, if forced down, I want protection from the sun for my arms.

The important point for all emergency equipment is to practice its use and be certain it's located in the airplane where you can get at it quickly. It's mandatory to simulate an emergency and actually try getting the equipment out and setting it up. That's the time and place to find out if it's awkward or impossible to get the life raft out a window— not after you've ditched and have only a minute to get it out before the airplane sinks.

With the equipment, and the assurance that you can get

at it, flying over the sea is quite relaxed, but I want emergency equipment even from Miami to Nassau, or Long Beach to Catalina Island.

Over-Sea Climate

Near coastlines, especially in northern and temperate areas, one finds fog more often than inland. Combatting it is a matter of having alternates inland where there isn't fog.

Otherwise, over-ocean weather isn't as violent as over land, with the exception of hurricanes, typhoons, or whatever you want to call them. Thunderstorms aren't as turbulent, although they may look awful. Aside from fog, a heavy shower at destination, or avoiding hurricanes, flying over the sea is an easier weather experience than from Chicago to Kansas City.

I like flying over the sea. The forced-landing risks are far less than pilot-navigation goofs or an improper letdown in stormy weather at the destination.

When Rob and I went to Australia in the 402, we spent many enjoyable hours watching the beautiful blue Pacific and its clusters of small cumulus clouds causing ever-changing shadows and color on the surface. As our destination island came near, we'd see ahead the more solid, taller clouds that tell you land is beneath. Before long the coast appeared, and the dark blue water changed to light shades of aquamarine as we flew across the sandy beach with its palm trees. It's romantic enough to make you think of Melville, Somerset Maugham, and the strains of "Sweet Lelani."

19

Flying Know-How

Jim Gannett is Boeing's senior test pilot, one of the finest and most respected in the world. It's been my good fortune to know and fly with him. We were part of the crew on the Rockwell Polar Flight, which was first around the world over both poles. We alternated the jobs of captain and copilot with three others: Fred Austin, Harrison Finch, and Jack Martin.

The start was Honolulu, nonstop to London over the north pole, then Lisbon, Portugal. After that Buenos Aires, and then across the south pole to Christchurch, New Zealand, and finish at Honolulu.

Out of Christchurch Jim flew captain and I copilot. It was a joy to watch him. Settled in the seat, he checked the items of preflight. His motions were sure, smooth, and at a pace that seemed slow but, because of their simplicity, really were quick. There never was a suggestion of being rushed, nervous, or dramatic. He was cool and deliberate.

In flight his management of the airplane was low-key, and when he gave an order it was in a gentle but positive voice, so you knew exactly what he wanted, and knew he knew what he wanted.

His flying was precise: best climb speed maintained exactly; when cruise altitude was reached he leveled off smoothly so you couldn't tell it; and he leveled exactly on the altitude, not five feet from it either way—he came around to and stopped exactly on the proper heading. You couldn't tell the controls were being moved, he anticipated so well and was so super smooth. He dominated the airplane; he flew it, it didn't fly him.

Watching Jim, you learn how easy flying is, how effortless this wonderful art can be. But how do Jim Gannett and his like accomplish this simplicity? They study and practice, enthusiastically pursuing the art of flight, until knowledge gives them confidence and practice provides the skill to perform.

It is there for all of us.

Index